Who Speaks for Nature?

WHO SPEAKS FOR NATURE?

On the Politics of Science

Laura Ephraim

PENN

UNIVERSITY OF PENNSYLVANIA PRESS

PHILADELPHIA

Published by
University of Pennsylvania Press
Philadelphia, Pennsylvania 19104-4112
www.upenn.edu/pennpress

Printed in the United States of America on acid-free paper

10 9 8 7 6 5 4 3 2 1

Library of Congress Cataloging-in-Publication Data

Names: Ephraim, Laura, author.
Title: Who speaks for nature? : on the politics of science / Laura Ephraim.
Description: 1st edition. | Philadelphia : University of Pennsylvania Press, [2018] | Includes bibliographical references and index.
Identifiers: LCCN 2017026854 | ISBN 9780812249811 (hardcover : alk. paper)
Subjects: LCSH: Science—Political aspects. | Science—Philosophy. | Political science—Philosophy. | Natural history—Philosophy. | Nature—Political aspects.
Classification: LCC Q175.5 .E64 2018 | DDC 303.48/3—dc23
LC record available at https://lccn.loc.gov/2017026854

CONTENTS

Introduction: The Science Question in Political Theory

When scientists speak publicly about things they have observed, measured, and analyzed in the lab or the field, their words carry special weight. Unlike more casual observers of the physical milieu—patients, mothers, poets, gardeners, laborers, cloud watchers—natural scientists are regarded as authorities when it comes to their objects of study. To acknowledge this much is not to deny that the authority of science and scientists can be challenged—for example, when citizens from Woburn, Massachusetts, created a "leukemia map" to refute experts' repeated claims that the toxic waste in their community was benign, or when members of ACT UP chanted "We're here to show defiance for what Harvard calls good science" while protesting the protocols and priorities of AIDS researchers.[1] But as such activists understand better than anyone, citizen movements face an uphill battle when they contest a scientific consensus. Today, the authority of the natural sciences—and the stakes of movements to support, co-opt, or erode that authority—are all the more visible in light of the twin dangers of climate change and climate denial. Global movements for environmental justice and sustainability seek to buttress the authority of the natural sciences, recognizing that this authority may be the last best hope in the struggle for earth's future against moneyed fossil-fuel interests.[2] Meanwhile, these interests spend lavishly to sow doubt about the climate science consensus, suggesting that they, too, recognize the authority of science as a formidable obstacle to their agenda.[3] If the question that frames this book, *Who Speaks for Nature?*, is seldom asked directly in these struggles, or in academic and political debates more generally, perhaps this is because the answer is obvious: scientists do.

But if scientists' capacity to speak authoritatively on behalf of the material environment seems almost self-evident, it is far less apparent how this authority comes to be established, sustained, and eroded, and under what conditions it is successfully vested in or wrested from the natural sciences. From where do scientific experts derive their enduring—though contestable—capacity to give voice to nature? How do communities come to be configured around the acknowledgment that (some) political actors (and not others) speak for the elements and energies of the material milieu and the flesh of human and nonhuman bodies? What practices on the parts of scientists and nonscientists establish and erode the boundaries circumscribing who counts as a legitimate spokesperson for which things, and whose speech is heard as idle chatter, "mere" opinion, or noise? This book sets out to dispel the aura of inevitability surrounding the authority of the natural sciences by interrogating the political origins, sources, and limits of that authority. In these pages, I elaborate a novel account of the political conditions of possibility that render certain natural phenomena speakable and that decide who may and may not speak on their behalf. By uncovering the political constitution of scientific authority, I aim to foster a deeper understanding of the contingency and the resiliency of that authority and, thereby, to inform and inspire ongoing movements to bolster, contest, or reshape it for democratic and ecological ends.

One obstacle to posing "the science question" in political theory in terms of the *political* constitution of scientific authority is the widespread assumption that scientists—unlike most political actors—primarily derive their authority from the truth of the things they say.[4] Within this epistemological framing, scientific authority is understood to be founded, in principle, on validity. In practice, this is usually taken to mean that scientists earn special trust from the public insofar as they restrict themselves to making claims about nature that have been verified to a high degree of certainty using a scientific method, that is, systematically testing hypotheses against carefully gathered evidence. We can see this epistemological interpretation of scientific authority at work, for example, when environmentalists cast their allegiance to science as an almost compulsory form of deference to overwhelming quantities of evidence; we even see it when climate deniers cast their reluctance to defer as a way of conforming with the scientific method's injunction to doubt and retest seeming certainties.[5] The tendency to conflate natural scientific authority with epistemological validity has also shaped generations of critical scholarship on the natural sciences in the humanities and the social sciences: Post-positivist philosophies of science have deflated confidence in the capacity of method to guarantee correspondence between scientists' linguistic claims and the reality to which they are supposed to correspond.[6] And in history, sociology, anthropology, and the interdisciplinary field of science and technology studies (STS), scholars have unveiled unseemly discrepancies between the ascetic prescriptions of the scientific method and the messy imbrications of actual natural scientists in patriarchal, racist, colonial, heteronormative, and otherwise troubling social and political relations.[7] The prevalence of these "debunking strategies," as Lorraine Daston calls them, reflects the widespread expectation that to debunk is to deauthorize—in other words, that the best way to defuse abusive versions of scientific authority is to reveal the irreducible, invalidating influence of interests, norms, and power upon scientific practices and discourses.[8]

Yet the authority of the natural sciences has largely survived these debunking gestures, suggesting that this authority cannot be adequately explained by epistemological considerations alone. Indeed, many of the very same scholars who embrace the post-positivist view that every scientific fact comes laden with values are *also* convinced of the importance of heeding and amplifying the voices of scientists when it comes to climate change. These positions only seem incongruous with one another if we assume that the authority of science stands or falls on the strength of its validity.[9] By moving beyond a narrow epistemological framing of this authority, we can better understand why the belief that all science is socially constructed usually does not—and should not—move believers to remove the Darwin fish decals (ubiquitous emblems of deference to science) from their fuel-efficient cars. Neither the healthy skepticism of post-positivism nor the cynical skepticism of climate denial has succeeded in eradicating natural science's authority, because this authority also draws from nonepistemological wells. These are the wells that interest me in this book.

Accordingly, I largely leave to one side questions about whether and how scientists know what they (claim to) know. These are important and engrossing questions, but they already take up too much oxygen in discussions of science and politics, and their prevalence unfortunately tends to incline even astute questioners to treat politics as little more than an invalidating dalliance. In place of the familiar epistemological questions, I open neglected lines of inquiry into whether and how political activity serves as an enabling condition for scientists to do what they do when they speak for nature. Beyond the methodological rigor of scientists' research, what habits, practices, disciplines, or comportments—on the part of those who do and do not have keys to the lab—render some efforts to give voice to nature more felicitous or legible than others? Though the question of whether scientists are speaking the truth is hardly irrelevant from either a political or an epistemological perspective, it would be naïve to think that scientists' capacity to serve in the crucial role of truth-teller depends solely on their access to truth.

Over the course of the next four chapters, I make the case that natural scientists owe their prevalent role as spokespersons for nature to what I will call the politics of world-building. By "world-building politics," I mean the embodied practices through which scientists and citizens struggle with and against each other to engage the material reality of their environments and bodies and compose a common world from these heterogeneous elements. World-building practices instantiate relations of proximity, affinity, resemblance, or repulsion among disparate human and nonhuman beings, excluding some from the assemblages that secure the power, prestige, and visibility of others. Seen within this worldly frame, the natural sciences are political because they are among the most important sites for inheriting, augmenting, dismantling, and rebuilding the material relationships that bind human and nonhuman bodies together and enable some bodies to speak for others. From this perspective, the political importance of the natural sciences is founded not only on what scientists know, but also on what they—and we—*do* in, to, and through the phenomena of nature. This worldly reframing of the politics of science encompasses within "the political" those practices that demarcate science as distinct from other modes of responding to matter and elevate the vision and voice of those who are counted as scientists above other citizens and subjects. The politics of world-building cannot be confined to the lab or the field; it involves both experts and nonexperts from many walks of life in founding and refounding authority for the natural sciences.

The idea that those who will defer to authority must be involved in producing it is a familiar trope from the social contract tradition. But unlike the extraordinary scenes of consent-giving portrayed by Hobbes and Locke, the decisive moments in the constitution of scientific authority portrayed in coming pages are quotidian exercises of a demotic, corporeal power: namely, common sense. Unlike traditional castings of common sense as a repository of conventional wisdom and a folksy epistemological foil to the sciences, this book redescribes common sense as a potent well of creativity and scientific authority's enabling condition of possibility.[10] Common sense, on this account, emerges when plural spectators perceive the same things from disparate vantage points, engendering for the objects of their shared scrutiny the palpable "realness" or "sheer thereness" (in Hannah Arendt's words) they would otherwise lack.[11] Common sense shoulders some bodies below the threshold of visibility and intelligibility and rescues others from worldlessness, granting them a meaningful, tangible place in the common world. Thus conceived, common sense plays a profound role in constituting scientific authority by determining both what in the world is perceptible and speakable and who is perceived as speaking for it. My account of common sense and the politics of world-building works with and against the legacy of the social contract tradition to reveal the interinvolvement of world-building and consent-giving, matter and meaning, and nature and politics.

The role of common sense and the politics of world-building in establishing natural scientific authority is obscured not only by epistemological traditions that affiliate common sense with prejudice and cast science as its overcoming, but also by ontological traditions that oppose the world of politics to the reality of nature. The tendency to treat nature and politics as though they were radically different orders of reality runs deep in the history of political thought: think of the Aristotelian contrast between animal pleasures and human justice, the traditional social contract narrative of covenanters extricating themselves from a prepolitical state of nature, or the Kantian opposition between material necessity and moral freedom. Dualisms between nature and politics, matter and meaning, and the given and the made make it difficult to recognize *as* political the manifold engagements with materiality that contribute to the constitution of a common world. Moreover, this dualistic style of thinking makes it difficult to recognize what scientists do to or say about natural phenomena as political practices in their own right, enabled by and consequential for the wider political world. The profound, longstanding influence of nature-politics dualities in political theory also helps to explain why so few political theorists have engaged deeply with debates in STS about the roles of politics in science and science in politics.[12] Virtually the only sustained discussion about science in political theory over the last half-century concerns the scientific status of *political* science. As we will see in the next section, the political stakes of *natural* science emerged only tangentially in this discussion, while the traditional foil between nature and politics was left unquestioned or even reinforced. In order to put the question of the politics of natural science back on the table in political theory, we must struggle to simultaneously interrogate epistemological oppositions between science and politics and ontological oppositions between nature and politics. It is not easy to do both things at once; I can only hope that this book and its inevitable shortcomings will be the beginning, not the end, of such efforts.

Fortunately, when it comes to the aim of contesting traditional nature-politics oppositions and imagining alternative ontological constellations, I am in increasingly good company. In recent years, a diverse movement to re-center affect, the body, and materiality in political theory has emerged from many corners—a movement I call "the worldly turn." A number of feminist, green, new materialist, aesthetic, and democratic political thinkers have contributed to the worldly turn. Their works rebuke traditional portrayals of material reality as an essence, cause, limit, or other to politics and recast political reality as the often unpredictable outcome of embodied, agonistic encounters among diverse human and nonhuman beings. The world these works turn us toward is at once material and meaningful, organic and artificial, corporeal and discursive. By revealing some of the many ways that "matter matters" (as Karen Barad puts it) in politics, the worldly turn makes it easier to raise the question of how the *sciences* of matter come to matter in politics.[13] Yet, with a few telling exceptions—and for reasons explored in a subsequent section below—very few of the thinkers involved in this worldly turn are prepared to recognize the natural sciences as worldly political phenomena in their own right. This book's account of the political origins of the authority of natural science fills some of these gaps in the worldly turn's rematerialized portrait of politics. To move more fully beyond specious nature-politics oppositions, we need to carefully scrutinize the central role of both natural phenomena and natural scientists in the constitution of the world of politics.

The framing question "Who speaks for nature?" responds to this need to push natural phenomena closer to the center of political analysis, but it does so by maintaining a distinction between the natural phenomena that are spoken for and the subjects who vie to speak for them. To some of the post-humanist thinkers affiliated with the worldly turn, this approach might seem to draw too firm a line between the mute object and the speaking subject—to center the non-human "that" at the cost of defining it in opposition to the human "who," and thus to succumb to the dualistic ontological attitude I just professed to resist. I would agree that the line separating non-speaking objects from speaking subjects is plastic, but not infinitely so; a variety of factors, ontology and history among them, conspire to render this line rigid in many places. To recognize this relative rigidity is to heighten, not discount, the importance of understanding the conditions under which political world-building practices may play a role in bending, maintaining, and sometimes dislocating the line between speaking and nonspeaking beings. Indeed, it may be misleading to picture anything so neat as a line there: between the terrain of the unambiguously speaking subject and that of the unambiguously mute object is a zone of ambiguity, itself of ambiguous size, populated by beings whose capacity for speech is subject to various degrees and kinds of doubt. Gayatri Chakravorty Spivak points to one important site within this ambiguous zone and highlights the difficulties of escaping it with the title of her famous essay, "Can the Subaltern Speak?"[14] She reminds us that colonial legacies continue to render the voices of many putatively human subjects barely recognizable as such, objectifying and dehumanizing them by confining them to this ambiguous gap between speech and silence (or noise). In light of Spivak's argument, we can see that the act of speaking *for* is a distinctive and deeply political form of boundary work. As Mark Reinhardt argues, those who claimed to speak for the fugitive slave Margaret Garner at once asserted their own political standing as human subjects and reinforced the voicelessness of the person they sought to "ventriloquize."[15] To seek to be heard as speaking for nature is a similar kind of boundary work—a way of shoring up the humanity and authority of some beings and confirming the classification of others as objects.

But there are also important, illuminating differences between the politics of speaking for the subaltern and the politics of speaking for nature. The "propensity to ventriloquize" that Reinhardt analyzes in the case of Garner—"putting words in her mouth and claiming access to her inner life"[16]—both presupposed and reinforced her ambiguous political positioning somewhere between subject and object. In other words, for the illusion of ventriloquism to have a chance of being a *successful* illusion, there must be a question mark hovering over the capacity of the ventriloquized to speak, as there is a question mark in the title and spirit of Spivak's essay. Whereas the voice and inner life of the beings categorized by Spivak as subaltern are *almost* unrecognizable in the archives of the dominant, the voice and inner life of the beings I categorize as natural are unambiguously unrecognizable. This lack of ambiguity is the reason the phenomena of nature, thus defined, cannot be ventriloquized when spoken for: there is nothing recognizable as an organ of speech to put words into, nothing recognizable as an inner life to give voice to. At least, not yet: none of this is to deny that ambiguity can sometimes be *created* and new question marks attached to beings that (who?) formerly seemed unambiguously voiceless.[17] Sometimes such ambiguity is created by the sciences, as it was by Project Nim, whose researchers sought to teach sign language to a chimp.[18] These scientists endeavored *not* to be heard as speaking for Nim Chimsky, as they called him; on the contrary, they wanted to show that chimps could speak for themselves, denaturalizing and even humanizing them by moving them into the zone of ambiguity. Sciences that seek to get nature to speak for itself are a fascinating rarity, but they are beyond the scope of this book. I prioritize analyzing sciences that seek to speak *for* nature because of their distinctive, undertheorized political implications: when such sciences succeed, they tend not only to reconfirm the classification of their objects of study *as* mute objects, but also to create or exacerbate doubts about the standing of the other speaking subjects who vie to be recognized as speaking for nature. This contest over who speaks for what is a dimension of the politics of science that is seldom appreciated as such, but comes into focus within the worldly frame of this study.

Coming chapters unfold this unusual conception of the politics of science by turning back to early, formative moments in the trajectory of the natural sciences, when their authority to speak for nature was new and more unsettled and open to radical transformation and reconstitution. I recover surprising insights into these constitutive moments from three early modern thinkers whose works confront and provoke profound uncertainty about who speaks for nature: René Descartes, Thomas Hobbes, and Giambattista Vico.[19] Descartes, Hobbes, and Vico may seem like unlikely allies to enlist in the cause of contesting ontological oppositions between nature and politics and complicating epistemological framings of scientific authority. Their works are commonly seen as typifying, if not originating, modern motifs opposing nature, as a mechanical domain of prediction and control, to politics, as an unruly domain of human activity and artifice. And Descartes, Hobbes, and Vico often present themselves as founding their respective versions of the "new science" on epistemological grounds, promising new and more reliable truths as the fruits of new scientific methods.

But the question remains: Where do they each muster the standing to gain a hearing for these disenchanted portrayals of natural matter? To whom do they promise their new truths, in exchange for what, and how do they propose to cement these obligations in practice? What political means do Descartes, Hobbes, Vico, and their followers and critics advocate and enact in the hope of defeating rival claims to speak for nature by the Academy, the Church, or (some iteration of) the people themselves? By approaching their texts as traces of struggles to challenge and refound the established grounds of scientific authority, I discover evidence of their participation in world-building practices that defy the ontological and epistemological logics that they sometimes affirm. While I don't doubt their faith in the truths they each espouse, I show that none of them depend upon validity alone to secure authority for their respective scientific projects. And I argue that, in their efforts to transform the political constitution and distribution of the authority to speak for nature, Descartes, Hobbes, and Vico each confront and contest ontological oppositions between nature and politics. In the context of their innovative efforts to authorize science, they each radically reimagine the materiality of politics and the politicality of science in ways that productively confound the familiar terms of discourse about natural science and politics today.

My approach to reading early modern scientific treatises for insights into the worldly political constitution of scientific authority emulates Hannah Arendt, whose critical engagements with the natural sciences are the focus of the next chapter. Arendt is well known for theorizing "the world" as the material home for politics and is already a source of inspiration to many in the contemporary worldly turn.[20] But her deep concerns about the worldly entailments of the natural sciences have gone largely unrecognized. Arendt genealogically redescribes the natural sciences—beginning with Galileo and moving through Descartes, Hobbes, and Vico—in political terms, as radical reorganizations of the corporeal, sensuous practices through which we encounter nature and build a common world. Her genealogical and worldly orientation to events of scientific transformation makes her keenly sensitive to a *problem* surrounding the natural sciences that other approaches do not equip us to see: namely, that emergent scientific regimes tend to put at risk the very world-building politics on which they also depend. Arendt helps me to raise the question of the relationship between natural science and common sense as a premier political question for the modern age.

But I repurpose Arendt's genealogical approach in ways that complicate her conclusion that the rise of modern science constitutes a tragedy for modern politics. I find reason to see both risk and promise in the authority of science with the help of Vico, one of Descartes's most trenchant eighteenth-century critics. Moving counter-chronologically from Arendt to Vico helps me to foreground both the dangerous and the emancipatory political potential of different modes of authorizing different kinds of sciences to speak for nature. When I then move further back in time to scrutinize Descartes's and Hobbes's respective efforts to vest authority in their respective new sciences in the seventeenth century, I do so with a mix of Arendtian trepidation and Vichian critical optimism.

As the temporal direction traveled by the book suggests, I do not use the texts of Arendt, Vico, Descartes, and Hobbes as stepping-stones in a traditional intellectual history. Arendt's and Vico's critical genealogies are not histories of science, and neither is mine. I prioritize intra- and intertextual interpretation ahead of contextualization and creatively appropriate works from the past to serve as vantage points from which to defamiliarize and reexamine the dynamics of the present. In this way, I recover from some of the most canonical works of the Western tradition an iconoclastic origin story, in which both the sciences and common sense play leading, vexing, and invaluable roles in struggles to remake the world and to make sense of its heterogeneous constitutive parts.

But first, in the remainder of this introduction, I consider in more detail the nature-politics binary that has constrained political theory's existing engagements with science as well as the promise and limits of existing challenges to that binary coming from the worldly turn. The next section traces the prevalence of what I call the "two-sciences settlement" across several generations in political theory. I argue that this oppositional view of nature and politics and the sciences that study them risks misdescribing political reality in the name of protecting it against scientific influence and misunderstands the grounds of the scientific authority it mourns. A third section then turns to a specific current within the worldly turn, the new materialisms, for help in challenging these ontological dualisms and rethinking the political significance of the material sciences. New materialist thinkers have been unusually bold in defying the two-sciences settlement and enlisting scientists to speak for matter in political theory and practice. But they stop short of asking how scientists become recognizable as spokespersons for nonhuman nature, eliding the political conditions of the authority they celebrate. Then, in the fourth section, I identify promising directions for rethinking the origins, grounds, and implications of scientific authority from works in the "aesthetic turn"—a movement that I resituate as an important part of the broader worldly turn. Finally, I briefly summarize the arguments to come in the book's remaining chapters.

The Two-Sciences Settlement

Over the last half century, political theorists seldom directly interrogated the political significance of the natural sciences. Instead, attitudes toward natural science in contemporary political theory were obliquely shaped by two waves of struggle over the influence of natural scientific methods within *political* science. In the postwar period, such political theory luminaries as Leo Strauss, Hannah Arendt, Sheldon Wolin, Jürgen Habermas, and Hanna Pitkin strongly opposed the behavioralist revolution in political science, a movement to model the discipline upon a stylized portrait of natural science as the embodiment of positivism. More recently, an email circulated by an anonymous Mr. Perestroika in 2000 ignited opposition to the American Political Science Association and its flagship journal for unduly privileging research that emulated the quantitative, predictive, and value-free principles ostensibly exemplified by natural science while devaluing qualitative, interpretive, and engaged research, particularly involving non-Western communities.[21] In each period, critics of the reigning natural-science ideal argued that movements to "naturalize" political science are premised on a false analogy between nature and politics. They argued that natural and social science should be separate but equal, operating according to distinct methodological imperatives dictated by their distinct objects of study. This two-sciences settlement has proven valuable when it comes to defending marginalized approaches to the study of politics. But it is also costly, entrenching an untenable opposition between nature and politics that obscures the political constitution of both natural and political science. To better understand the sources and to critically analyze the effects of the authority of natural science—in political science and in politics—we must move beyond the two-sciences settlement and the traditional, two-worlds ontology that underwrites it.

Oppositions between nature and politics figured prominently in each of the two waves of *Methodenstreiten* in political science. Wolin's landmark 1969 essay "Political Theory as a Vocation" put it plainly: "Method is not a thing for all worlds."[22] In his view, methods suited to revealing the natural world's calculable regularities and deterministic laws are inapplicable to the political world's normatively weighty and contextually dependent phenomena.[23] Thirty-two years later, Perestroikan Mark S. Kremer invoked a similar contrast between nature's mechanical laws and the contingency and particularity of the laws governing politics: "Physics could only attempt to mathematize motion on the basis that the bodies in motion are not conditioned by anything other than physical laws of nature. Man, however, is conditioned by human laws or conventions."[24] Another Perestroikan, Gregory Kasza, further reinforced the idea that objects dictate methods: "Specialization may befit the natural sciences, but their objects of study are distinct."[25] For Kasza and Kremer, as for Wolin before them, those who would apply scientific methods to politics wrongly disregard the ontological differences between natural matter and the contextual meanings and subjective attitudes of politics.

Such ontological appeals for two separate kinds of science are indebted to philosophical defenses of the autonomy of the *Geisteswissenschaften*, the human or social sciences, from the *Naturwissenschaften*, or natural sciences, dating back to Wilhelm Dilthey. His 1883 *Introduction to the Human Sciences* defined "the fundamental problem" of the human sciences as the need for boundaries "establishing the precise kind of incommensurability between relations of intellectual processes and uniformities of material processes which would preclude reducing the former to mere characteristics or facets of matter."[26] For Dilthey, the proper purview of the human sciences is a realm of concepts, purposes, values, and experiences circumscribed by the borders of the human mind—an intellectual reality that he considered ontologically irreducible to the material reality studied by the natural sciences. This dematerialized view of the objects of the human sciences was reinforced in the twentieth century by the so-called "linguistic turn," as philosophers emphasized the role of language in constituting, not merely reflecting or describing, reality. For example, inspired by Wittgenstein's ordinary language philosophy, Peter Winch argued in 1958 that there is "a difference in kind" between social reality, which consists of conventional, context-specific rules for using and understanding language, and nature, which consists of matter governed by universal, causal laws.[27] Winch called for scholars of the linguistic version of reality to eschew the predictive goals of the natural sciences and embrace interpretation instead. Similarly, Hans-Georg Gadamer argued that linguistic understanding is both the "primordial mode of being of human life itself" and the mode of inquiry best suited to advancing knowledge of this mode of being.[28] The tradition of Dilthey, Winch, and Gadamer offered political theorists an appealing vocabulary with which to defend the legitimacy of interpretive, contextual approaches in political science against behavioralist attacks. But locating politics within an immaterial world of *Geist* occluded the various ways that both matter and the sciences that study it are themselves political phenomena.

To their credit, some postwar critics of naturalized political science nonetheless made strides toward analyzing natural science as a potent influence on both political science *and* political reality. For example, Wolin projected a coming "world of bleak, forbidding, almost sterile reality" as the growing authority of natural science served to "impart regularity and predictability" to politics.[29] Similarly, Arendt warned that scientific methods, mentalities, and experts threatened to turn politics into the very mechanical, homogenized, nature-like reality that behavioralists assumed it to be: "Statistical uniformity is by no means a harmless scientific ideal; it is the no longer secret political ideal of a society which . . . is at peace with the scientific outlook inherent in its very existence."[30] Indeed, as we will soon see, Arendt both marshaled and challenged the separate-sphere mentality of the two-sciences settlement tradition and went beyond Wolin to attend to the political *constitution* of the detrimental (to her and Wolin both) influence of scientific practices on the political world. In different ways, Arendt and Wolin were each concerned with the threat of the rationalist, instrumental ethos that they each associated with natural science to both political life and political science. They called for political science to resist the temptation to become a natural science as part of a broader struggle against the scientization of political *and* intellectual affairs.

The terms of the naturalist controversy had shifted by the time Mr. Perestroika distributed his 2000 email, reflecting two broad changes in the conventional meaning of "science" near the turn of the twentieth century. First, science's public reputation, which had suffered by association with Cold War nightmares of nuclear winter and iron cages, was increasingly rehabilitated in liberal circles as late twentieth-century movements turned to science with the hope of solving such nightmares as global warming and religious fundamentalism.[31] Reflecting this evaluative shift, Perestroikans were seldom as troubled by scientized politics as Wolin and Arendt. More attuned to the promise than the risks of public science, Perestoikans called for political scientists to become public-minded problem solvers.[32] But to become, in Bent Flyvbjerg's words, "a social science that matters," Perestroikans argued that political science must embrace different and more varied approaches to research than the natural sciences.[33] For example, Sanford F. Schram called for "a political science that forgoes the dream of a science of politics in order to dedicate itself to enhancing the critical capacity of people to practice a politics."[34] To become practically relevant, he argued, the discipline must dispel "the idea that the social sciences need to ape the natural sciences in the pursuit of scientifically tested and validated generalizations about reality."[35] Second, by the time of Mr. Perestroika's email, the dominant view of natural science among political scientists had shifted under the growing influence of post-positivism.[36] Political scientists were increasingly aware of Thomas Kuhn's redescription of the natural sciences as linguistically mediated social formations that move through incommensurable, value-laden periods of "normal" science.[37] The influence of Kuhn and other post-positivist epistemologies made it harder for Perestroikans to assume—as both sides in the postwar behavioralism debates had, by and large—that the natural sciences actually approximate positivist ideals of value freedom, objectivity, and universality.

It might seem as though post-positivism should have undermined the two-sciences settlement by leveling many of the supposed differences between the natural and social sciences, revealing them to be similarly situated, hermeneutic, and imbued with value commitments. But instead, Perestroikans reshaped the two-sciences settlement to better conform with Kuhnian insights. As Kasza characterized it, "The Perestroika movement is a reaction against scholars who wish to turn the study of politics into what Thomas Kuhn called a 'normal science'" by enforcing unity around a single, quantitative research paradigm.[38] To advance the argument that paradigms are only possible in the natural sciences, Perestroikans drew new ink from old ontological wells: they granted that the *subjectivities* of all scientists are steeped in contextual values and linguistic meanings, but contended that political *objects* are contextually and linguistically dependent, while natural objects are not. Flyvbjerg, whose work was widely influential among Perestroikans, explained the difference as follows: "Ideal natural science explains and predicts in terms of context-independent elements which can be abstracted from the everyday world—mass and position in physics, for example."[39] But the "'object world'" of the social sciences "is a subject world," its very elements derived from subjective experience and determined by context.[40] Schram concurred: "Regardless of the fact that both natural and social science are forms of learning in context that produce value-laden facts, social life, as opposed to the objects of natural scientific inquiry, involves multiple interpretive lenses that offer a cacophony of competing perspectives emanating from its origins in conscious, thinking human beings."[41] Schram, Kasza, and Flyvbjerg accepted the post-positivist insight that natural scientists and social scientists alike are social creatures, equally incapable of fully extricating their thoughts from their contexts. But if all theories and facts are value-laden linguistic constructs, they insisted, only some objects are. This revised two-sciences settlement differentiates between the objective objects of natural science and the subjective objects of social science (including, for Schram and Kasza, politics). Normalization around a paradigm, they argued, is only viable in sciences whose objects are free of the values and meanings that inevitably seep into the minds of the scientists involved. Political scientists, whose objects *are* subjects and their intentions, beliefs, and ideas, can only suffer from the hegemony of a single method. In Lloyd I. Rudolph's words, "Let a Hundred Flowers Bloom, Let a Hundred Schools of Thought Contend."[42]

This revised two-worlds ontology continues to have wide currency in political theory. For example, Keith Topper's recent friendly critique of Perestroikan methodological pluralism invokes Charles Taylor to argue for the "doubly hermeneutic" character of the social sciences: "Investigations of human phenomena are not just hermeneutic but doubly hermeneutic in the sense that they must take into account not only the interpretations of those engaged in the investigation—this, [Taylor] maintains, is equally true of the natural sciences—but also the self-interpretations of those being investigated."[43] In Topper's mapping, all sciences are hermeneutic, because all scientists are self-reflexive subjects. But only the social sciences are doubly hermeneutic: "The objects of social scientific inquiry are self-interpreting (individual and collective) agents and their meaningful actions, practices, norms, intentions, and institutions, rather than 'meaningless' events, processes, states, and behavior."[44] On such an analysis, reality cleaves into two segments, one context-dependent, the other context-independent; one intelligible by virtue of the human capacity to interpret, the other intelligible by virtue of methods that mitigate (though they cannot solve) the human tendency to project interpretations where they don't belong. Topper, like Schram, locates politics on the social-object side of the ontological cleavage and cautions on this basis against emulating the natural sciences in political science.

It is easy to understand why the theme of ontological differences between nature and politics would recur in response to repeated efforts to exclude from political science approaches that differ from what is perceived to count as worthy in the natural sciences. Yet there are reasons to question the picture of two worlds bequeathed to us by this history. Elsewhere, I have raised the question of how well this dualistic ontology serves even the Perestroikan cause of methodological pluralism: if we accept the idea that ontology dictates methodology, it is difficult to understand why a hundred schools of thought should not cede to the few (or the one) most precisely tailored to the ontological contours of politics.[45] Against the wishes of its advocates, the two-sciences settlement unduly constrains political scientists' self-understanding of their discipline. Here I would stress that this settlement and its underlying dualist ontology also constrain our understanding of *politics*, obscuring the political significance of both nature and the scientists who study it. The picture of two worlds, *Geist* versus *Natur*, drawn by Dilthey and revised over generations, improperly restricts which institutions, formations, controversies, events, and actions can count as political, blinding us to the hybridity of some of the most pressing issues we face today, such as financial bubbles, drone strikes, oil spills, and water rights. Consider an event like 2005's Hurricane Katrina. The two-sciences settlement allows us to acknowledge as political the contested meanings of this storm—for example, shifts in public opinion about the causes of extreme weather or contending discursive constructions of race in media depictions of predominantly black and poor people taking shelter. But this settlement requires us to bracket as "immaterial" to politics the matter *of* or *about* which these avowedly political views, ideas, agendas, significations, and interpretations are constructed: for example, the storm's wind and water, the toppled concrete walls, or the bodies of humans and other living beings in their precarity and resilience.[46]

But physical matter and political meaning cannot be divided in this way without doing violence to reality. This is not to say that Flyvbjerg, Schram, or Topper would be wrong to paint the child sheltering in the Superdome or the president flying overhead aboard Air Force One as self-interpreting subjects, or to entertain serious doubts about the self-reflexivity of wind or water. But their underlying interpretation of interpretations—what they are and why they matter—is questionable. They treat interpretations as subjective stuff—dimensionless, weightless, and invisible—"emanating from" human consciousness. Such a view dematerializes politics by interiorizing it, secreting "the political" within the worldless cavern of Dilthey's *Geist*, where ideas, agendas, understandings, and judgments are ostensibly formed and stored. This dematerialized interpretation of interpretation artificially separates the corporeal event of encountering or avoiding Katrina's material manifestations from the judgments formed by displaced New Orleanians and their absentee president. It elides the origination of their contending interpretations in and through their lived, physical experiences of this extreme weather event and its aftermath. Together, the two-sciences settlement and the linguistic turn discount the materiality of meaning itself, and thus of politics. Although Perestroikans accord the status of "reality" to the interpretations analyzed by political scientists, they stop short of treating the hermeneutic stuff of politics as a real, constitutive part of the material world, including the material stuff of nature studied by the natural sciences.

This book derives resources from Arendt, Vico, Descartes, and Hobbes to resist the idea that interpretations, symbols, stories, images, and languages exist on a different ontological plane than material reality. It offers an alternative account of meaning as an integral part of what Arendt calls the worldly "in-between," a reality created when human and nonhuman bodies are moved into relation by some combination of intention, inspiration, accident, and gravity. Making sense of things is a way of remaking reality—a world-building practice every bit as physical as the practice of building houses. To return to the example of extreme weather, human beings contribute more to the climate than just carbon dioxide. Our world-building politics contribute to the environment a network of corporeal relationships, binding disparate elements together and leaving others adrift and worldless. Meaning, in this view, is not an overlay upon matter but its interstitial tissue; politics is not a machine for transmitting immaterial ideas to multiple subjective outposts, but a loom upon which the extrasubjective fabric of the world is woven. Without politics, there would be no world, but only "a heap of unrelated articles, a non-world," in Arendt's words.[47] Through the politics of world-building, we add relation, measure, and number to a motley collection of articles, unheaping them to cohere with our own bodies into a world, a contingent solidarity that is greater than the sum of its parts. The objects of both natural and political science are to be found in the dimensional spaces between and among human subjects, not within the interior reaches of their minds.

This alternative vision of a single world, at once meaningful and material, presses us to reconsider not only the relationship between political science and natural science, but also the relationship between natural science and politics. Natural science is political not merely because scientists view their objects through subjective interpretive lenses, but also because these objects are themselves bearers of meaning and products of world-building practices. Interpretations are as integral to the material reality of natural phenomena as their atomic number or cellular structure. Natural scientists themselves contribute profoundly to the political activity of reworking material configurations of meaning in the course of their research—at times building upon the work accomplished outside the lab or the field, and at times disdaining or destroying it. If there is only one world, at once meaningful and material, then the objects studied by the material sciences cannot be fully isolated from the meaningful configurations into which political activity—including the activity of science—sets them. Indeed, any effort to neatly divide the material practices of scientists from the material practices of nonscientists is itself a political move, a way of severing some of the strands of the world's relational web and knotting others together more tightly. In this respect, the two-sciences settlement unwittingly participates in a kind of politics that it also works to erase: the worldly politics of organizing and contesting material configurations of human and nonhuman beings and determining who may say what about which of the world's heterogeneous elements.

The New Materialisms and the Authority of Science

Thinkers affiliated with the new materialisms share my refusal to consign all things hefty or fleshy to an ostensibly apolitical realm of nature.[48] But where I have been emphasizing the materiality of human political activity, new materialists more often accent the politicality of nonhuman matter. Frustrated with ubiquitous renderings of matter as passive, inert, dull, deterministic, or mechanical, they seek to reveal the spontaneity, creativity, power, and freedom of nonhuman things, assemblages, and flows. The new materialisms offer innovative portraits of matter as an effervescent pooling of imminent agency, capable of enfolding itself into all manner of quasi-purposive "actants"—a term that emerged in the actor-network theory of Bruno Latour but that has taken on new life in the new materialisms. Jane Bennett explains: "An actant is a source of action that can be either human or nonhuman; it is that which has efficacy, can *do* things, has sufficient coherence to make a difference, produce effects, alter the course of events."[49] Such redescriptions of matter as vibrant, lively, and agentive help to pluralize the "onto-stories"[50] at our disposal in political theory and discourage the bad habit—evident not only in the two-sciences settlement—of treating the billiard-ball motifs of classical physics as unassailable ontological dogma. But there is reason for concern that the versions of posthumanism that have emerged from the new materialisms often obscure the political origins and consequences of the special authority of one class of human actants—natural scientists—to speak for nonhumans.

By the standards of political theory, new materialists are unusually eager, appreciative readers of natural scientific research. William Connolly, Samantha

Frost, John Protevi, and Brian Massumi each devote considerable effort to educating themselves and their readers about discoveries in neuroscience, biology, particle physics, and chaos theory. This research, they show, unexpectedly redescribes matter in terms that violate ingrained Newtonian expectations, informing their own accounts of thingly agency. Connolly has been at the forefront of this turn to natural science in political theory since 2002's *Neuropolitics*, which draws upon chemist and chaos theorist Ilya Prigogine and neuroscientists Antonio Damasio and Joseph LeDoux, among others, to "teach us a thing or two about the layered character of culture" and its imbrications with "a nature that is sometimes creative and novelty producing."[51] Similarly, Coole and Frost offer primers on recent developments in physics and biology in the introduction to their coedited *New Materialisms*; these contemporary sciences envisage "a considerably more indeterminate and complex choreography of matter than early modern technology and practice allowed," they argue, "thus reinforcing new materialist views that the whole edifice of modern ontology regarding notions of change, causality, agency, time, and space needs rethinking."[52] Frost recently took the extraordinary step of pursuing graduate study in the life sciences. She speaks with expertise when she describes the matter of human bodies (and, by extension, the politics in which those bodies participate) in terms of the constrained energetic relations of the cell wall's molecular fabric.[53] By defying the informal travel ban that kept generations of political theorists from visiting the journals and labs of natural scientists, Frost, Connolly, and other new materialists help to erode the two-sciences settlement and promote healthy curiosity about the insights into politics available through close readings of natural scientific discourses.

Yet they resist the suggestion that the natural sciences should be approached *as* discourses. With reason, they associate the view of science as a discursive construction with the linguistic turn, and they consider this emphasis on language an impediment to theorizing matter on its own terms, apart from human inscription. "Language has been granted too much power," writes the feminist theorist and physicist Karen Barad, in this vein. "Language matters. Discourse matters. Culture matters. There is an important sense in which the only thing that does not seem to matter anymore is matter"—a situation she redresses by guiding her readers through the complexities of Niels Bohr's quantum physics.[54] Similarly, Coole and Frost consider "the dominant constructivist orientation to social analysis . . . inadequate for thinking about matter, materiality, and politics" and look to natural science to help political theory past its "allergy to 'the real' . . . whereby overtures to material reality are dismissed as an insidious foundationalism."[55] John Protevi shares their sense that "the real," as it is disclosed by contemporary physics and biology, is both more realistic and less politically insidious than the versions of reality offered by natural scientists in the early modern past: "The acceptance of certain innate features need not underplay the political when we remember the emphasis critical biology places on the life cycle, developmental plasticity, and environmental co-constitution."[56] Protevi proposes that it is finally possible to be earnest, not strategic, biological essentialists, for the bodies described by the most up-to-date biologists are essentially spontaneous, not deterministic; agentive, not necessitated; and queer, not dimorphic. Together, these new materialists suggest that we no longer need the linguistic turn to protect politics from essentialism, determinism, and foundationalism because, at least since Darwin and Einstein, the natural sciences have revealed the material foundations of politics to be unpredictable, agentive stuff.

New materialists are right to criticize versions of the linguistic turn that neglect the materiality of politics, and their openness to the unfamiliar onto-stories relayed by contemporary natural scientists is refreshing in political theory. But the terms of the friendship they would broker between the natural sciences and political theory are questionable. When new materialists look to the sciences to speak for nonhuman actants like quarks and DNA, they seldom scrutinize the scientists doing the speaking as agentive political beings in their own right or interrogate the material political conditions that underwrite this authority. The capacity of scientific discourses to enable and constrain political reality is just one of the dimensions of the politics of science that deserves renewed—not diminished—attention as part of a thoroughgoing critique of the linguistic turn and reappraisal of the relationship between meaning and matter. Connolly, Frost, and Protevi each acknowledge that scientific discourses shape prevalent cultural motifs or political beliefs, but they seldom stop to analyze this aspect of scientific authority or explain its implications for a posthumanist, antidualist politics or political theory. Without greater sensitivity to the political origins and outcomes of the authority exercised by Bohr or Prigogine, we risk distorting their postclassical onto-stories into next-generation ontological dogmas, narrowing instead of widening the stories that can be told about how matter matters in politics.

We also need to pluralize the stories that can be told about how *science* matters in politics, drawing connections between what scientists do with language (itself a material phenomenon) and what they do with their hands, eyes, specimens, money, and instruments. As Ian Hacking rightly emphasizes, scientific practices of representing and intervening in material reality are inextricable from one another.[57] The new materialisms help to create the opportunity to build on Hacking and illuminate the political underpinnings and consequences of natural scientists' interventions into their objects of study. Yet even the political theorists who have been most eager to engage the natural sciences to promote new materialist outlooks have not pursued this opening to cultivate new, materialist political theories of the natural sciences themselves. This book helps to address this deficit by exploring scientific representations and interventions in tandem, refusing to dematerialize either language or politics. With the help of Arendt, Vico, Descartes, and Hobbes, I theorize the politics of science in worldly terms, as a constructive, creative activity of reconfiguring material reality.

In so doing, I work with and against Latour and the actor-network theory he, Michel Callon, and John Law developed in the 1980s and 1990s—an important minoritarian tradition in STS and a precursor to today's new materialisms.[58] Actor-network theorists conceive of science as a practice of assembling coalitions among researchers, the things they study, and the surrounding communities they depend upon and serve. In Latour's classic example, Louis Pasteur created alliances among technicians, veterinarians, farm animals, bacilli, and laboratory equipment in order to muster the agency to move the first vaccines into the bodies of France's human and livestock populations.[59] Latour emphasizes that Pasteur did not act alone; rather, his actions depended upon a network of relationships among diverse human and nonhuman beings, who likewise participated in and achieved agency through this network. Latour's concept of the "actant" thus treats agency not as an ontological given of matter, but as a provisional consequence of the organization of diverse material forces. As influential as Latour has been in encouraging political thinkers to center materiality in politics, his concurrent attention to scientists' own political agency, its enabling conditions, and its diverse effects has not received the attention it deserves. My approach revives this aspect of the Latourian project, attending to the practices that draw scientists, nonscientists, and nonhumans together into heterogeneous assemblages and distribute the capacity to speak and act unequally among them.

But I resist Latour's recent shift from describing the origins of scientists' agentive positions within actor-networks to prescribing their authority as a new norm for posthumanist politics. In *Politics of Nature*, Latour proposes a "new constitution" in which nonhumans would participate as equals alongside humans. Latour's constitutional ideal is a representative democracy, and he reserves for scientists the role of "spokespersons of the nonhumans."[60] "Now, who is better able than scientists to make the world speak, write, hold forth?" he asks, before affirming the answer his rhetorical question already cast as correct: "So, the sciences are going to put into the common basket their skills, their ability to provide instruments and equipment, their capacity to record and listen to the swarming of different imperceptible propositions [Latour's term for proto-actants] that demand to be taken into account."[61] Latour treats the fact that scientists are already recognizable as qualified to speak for nature as an argument in favor of enshrining this authority as a constitutional right. Because he stops short of asking how particular qualifications became recognizable as marks of authority, he naturalizes this authority, rendering its political origins and sources invisible.

Latour's turn to natural scientists to secure the inclusion of nonhumans in politics exemplifies what Nancy Fraser, in a different context, calls "the second dogma of egalitarianism": the view that the boundaries of political membership can and should be drawn by experts, circumventing the need for political struggles over who (or what) belongs.[62] Fraser observes this dogma at work when theorists of global justice look to empirical social science to measure the exact extent and intensity of transnational entanglements and thereby determine who owes what to whom in the name of justice. This way of globalizing justice, she argues, promotes an inclusive, democratically expansive answer to the question of *who* is owed justice, but at the expense of entrenching an exclusive, democratically narrow answer to the question of *how* to decide who is so owed. "The effect," Fraser writes, "is to neglect the importance of public autonomy, the freedom of associated social actors to participate with one another in framing the norms that bind them. Insofar as this approach confers the authority to determine the frame on social-scientific experts, it denies the public autonomy of those whom it subjects to those experts' determinations."[63] Analogously, Latour's move to confer authority upon natural-scientific experts presumes that it is possible—and desirable—to depoliticize the framing of representation. He would settle the crucial issue of *how* to decide *who* to include within the boundaries of the political by constitutional fiat: Scientific experts are to be responsible for deciding which among the "swarms" of nonhumans will be accorded representation and have their "demand[s] to be taken into account" satisfied.

In the name of establishing equality between nonhumans and humans, Latour naturalizes existing inequalities between those humans who are easily recognizable as spokespersons for nature and those whose voices would be discounted if they tried to inhabit this role. Because Latour leaves the origins of the scientific authority he affirms unexplained, he risks miscasting it as an inevitability rather than a contingent political accomplishment.

Latour's exercise in constitutional design makes explicit the authority that other contributors to the new materialist literature implicitly grant to natural scientists; for example, Barad's turn to Bohr or Connolly's turn to Prigogine as spokespersons for matter. They look to these natural scientists to lend authority to their redescriptions of nature as vibrant and agentive, and to erode the authority of competing descriptions of matter as dumb and passive—the still-hegemonic billiard-ball model that continues to derive *its* authority from earlier generations of scientists. But Barad, Connolly, and other new materialists too often leave unquestioned *how* natural scientists attained the authority to which they defer. When they turn to the sciences to speak authoritatively to political theorists about natural phenomena, they participate in material political dynamics that remain unthematized within their work.

Mark Brown helps to fill in these gaps in *Science in Democracy*, in which he attends to some of the political processes, institutions, and struggles that enable scientific authorities to play the role of representative of nature in politics. Brown—himself working with and against Latour—points out that the capacity of scientists to speak for nonhuman objects of study in politically meaningful ways is continually reshaped by other struggles over political representation; for example, the more commonly recognized politics of representation that determines who may act as a representative for which human constituency. Brown's "conception of political and scientific representation as practices of mediation that engage and transform what they represent" moves the science question in political theory in a productive direction and helps to open the way for this book's exploration of the political constitution of scientific authority.[64] But even as Brown highlights the role of human activities and institutions in mediating scientists' authority to represent nature, he leaves uninterrogated Latour's key assumption that it is scientists who are best equipped to serve this representative function when it comes to nonhuman constituencies. Indeed, Brown wants us to see that laypersons play a profound role in sustaining the capacity of scientists to speak for nature so that more of us will be willing to take up this supportive role: "The best way of responding to politicized science is to democratize it" by widening the human constituency willing to stand up for scientists' legitimate role in speaking for nature.[65]

This book renders the political significance of such deference explicit, not to condemn or endorse it, but to investigate it as among the enduring ambivalences that define the character of modern science and modern politics. By inquiring into the origins, promise, and limits of the distinctive standing of scientists in worldly networks of human and nonhuman beings, we will become better equipped to notice hitherto-neglected points of affinity *and* conflict between the practices of experts and laypersons. Like Latour, I approach science as a material praxis, but unlike him I follow Fraser's cues to critically analyze *how* the boundaries that circumscribe *who* may speak for nature are erected, assailed, eroded, bolstered, or reshaped. How did the capacity to intervene in and represent nonhuman matter come to be tied so closely to membership in a scientific discipline and access to the instruments and training these disciplines organize? Who is allowed or encouraged to access the skills that are cultivated in laboratories and other rarefied research spaces, and under what conditions do the alliances that these insiders forge with nonhuman beings empower or disempower laypersons beyond the laboratory walls? Through what course of events did the answer to Latour's question "Who is better able than scientists to make the world speak?" come to seem obvious? What configuration of matter and meaning makes it easier, by comparison, to discount the voices of nonscientists concerning the material things they struggle with and against day to day? Arendt, Vico, Descartes, and Hobbes are helpful in addressing these questions because they each grappled with the question of how the right to represent nonhuman nature has been claimed and disputed. Their respective efforts to contest who may claim the privilege of science illuminate the origins of partitions between those who are and are not seen as qualified to intervene into, ally with, struggle against, and represent the world's human and nonhuman parts.

A Worldly Aesthetic Turn

When I speak of "the politics of science," then, I have in mind the tumults, performances, comportments, and disciplines that both establish and sever relationships among various human and nonhuman beings, enabling some of them to speak on behalf of others. This worldly reframing of the science question in political theory owes a debt to Jacques Rancière's definition of politics as the contest over whose voices count and what they are perceived as competent to speak about: "Politics is not primarily a matter of laws and constitutions. Rather, it is a matter of configuring the sensible texture of the community for which those laws and constitutions make sense. What objects are common? What subjects are included in the community? Which subjects are able to see and voice what is common?"[66] Rancière's understanding of politics suggests that a juridical order, such as Latour's constitution, will only "make sense" if the community that it is supposed to govern has already established who is—and who is not—"able to see and voice what is common." Like Fraser, Rancière pushes us to interrogate *how* the boundaries and allegiances that delimit and texture communities are drawn. His conception of politics as "dissensus about . . . who is qualified to see or say what is given" suggests the need to attend closely to the reorganization of the aesthetic dimensions of established orders—the visibility of bodies, the audibility of voices—in movements to establish or transform scientific authority.[67] How is it that communities come to be sensuously configured around the perception that scientists speak for nonhuman nature?

The aesthetic turn in political theory, including Rancière's influential works, encourages a worldly orientation to politics by insisting on the creative, disruptive potential of the body's affective responses to its material milieu. Like Rancière, thinkers of political aesthetics such as Kennan Ferguson, Jason Frank, Nikolas Kompridis, Davide Panagia, and Linda Zerilli show how sensation, feeling, and other corporeal modes of experience are integral to the constitution of the common world.[68] By looking to the fleshiest dimensions of the human condition to explain the most transformational possibilities of politics, they implicitly challenge the two-worlds ontology on which the two-sciences settlement rests. But questions about the politics of science rarely surface in the aesthetic turn in political theory. This book builds upon aesthetic rethinkings of politics to recast the natural sciences as a series of interventions into the sensuous disclosure of the world. My worldly approach to the politics of science is informed by and extends the insights of the aesthetic turn in three main respects.

First, like others involved in the aesthetic turn, I approach the body's responsiveness to its physical surroundings as a wellspring of contingency, disruption, and creativity, rather than a site where causal determinism seeps into and compromises the freedom of politics. Panagia, for example, describes the power of sensation to disrupt customary ways of inhabiting the world, loosening existing political attachments and creating new ones. Moments of sensory interruption "are political moments," he explains, "because they invite occasions and actions for reconfiguring our associational lives."[69] The sense organs of the body, on such a view, are not merely passive receptacles for the raw material of subjective experience, but active, creative powers, capable of drawing hitherto imperceptible phenomena above the threshold of perception and organizing unprecedented solidarities. Melissa Orlie, drawing upon Aldo Leopold and Nietzsche, makes a similar point about the inaugural potential of affective experiences of nature. Pleasure, pain, and other passions are not automatic responses to natural phenomena, but creative ways of reworking reality, as epitomized by the aesthetic pleasure of beauty: "To see beauty is to *make* beautiful... rather than anything simply given in the nature of things."[70] Zerilli would concur: "Nothing compels us" to take pleasure in the things we judge beautiful. In aesthetic and political judgments alike, "'we feel our freedom,' as Kant put it . . . and discover the nature and limits of what we hold in common."[71] Zerilli, Orlie, and Panagia each treat aesthetic experience as an embodied, creative mode of at once engaging and transforming the material world.

The story of common sense as a world-building power that threads through this book extends this line of thinking about the creativity of the senses. Common sense, as I use this term, names a power that emerges when plural spectators sense one another in the act of sensing the things between them. Where a single body exercises its five private senses in tandem, the "sixth sense" of common sense arises from the organization of multiple bodies within a shared context. Common sense enhances the "realness" of the world, to use Arendt's word, and establishes embodied relationships among spectators occupying different vantage points. Arendt, Vico, Descartes, and Hobbes each help me to theorize common sense as a collectively enacted power to generate the sensuous texture of the common world. By delimiting which subjects may see and speak for which objects, common sense establishes and transforms the authority of scientists to speak for the things of nature.

Second, I follow Rancière's example in treating matters that may seem to be ontologically fixed—for example, which kinds of beings are perceptible and which are capable of speech—as contingent outcomes of politics. For Rancière, politics renders controversial who or what belongs to the common world and in what capacity. By maintaining a principled neutrality about the ostensibly ontological question of what is really there in the world, he converts it into a political question: Under what conditions may reality be reshuffled, enabling beings who had long been silent or silenced to step forward to speak and be heard? A similarly principled neutrality is merited with regard to the related, ostensibly epistemological question of who is best able to speak for nature, and on the basis of what criteria. Extending this style of questioning to the authority of science, I describe several instances when political dissensus redrew the very distinction between participants in politics and objects of science. These are the political dimensions of science that Latour risks suppressing when he reaffirms the unique qualifications of those who already count as scientists without asking how they became recognizable as such. With Rancière, I argue that politics renders newly contestable at one and the same time what is really there in the world and "who is best able" to draw it forward, make it visible, or represent it.

Maintaining an ontologically agnostic perspective on the conjoined questions of what the world is made of and who may speak for it need not stymie valuable efforts to re-center nonhumans in political theory. Like the new materialisms, the aesthetic turn has drawn attention to the inextricable political relations among human and nonhuman bodies: for example, Panagia's account of the chocolatiers, chocolate, and other "heterological elements" that converged in opposition to EU food standards, and Orlie's Leopoldian affirmation of movements to "compose a 'biotic complex'" inclusive of all life-forms, soils, and waters.[72] Somewhat like actor-network theory, these theorists underscore that it is difficult to predict in advance how matter will matter in politics, because the knots that enmesh diverse human and nonhuman bodies in a common world are contingent products of political struggles. Indeed, as Rancière emphasizes, one of the most unpredictable contingencies of political life is the question of which of the world's heterological elements will come to count as human, speaking subjects, which will be consigned to the status of voiceless objects, and which will be pushed below the threshold of perceptibility altogether. Collating the insights of Rancière and Latour helps me to show how movements to refound the natural sciences have been among the most important ways that alliances traversing and defining the line between humanity and its others are reconstituted.

Third, the aesthetic turn offers an alternative to the linguistic turn's dematerialized view of language by lavishing attention on speech as an event or scene where language is embodied and enacted. In rhetorical events of speech and debate, language lives in the body: a mouth moves, usually accompanied by the hands or feet, and this vibrates and sometimes stirs other bodies in the vicinity. As Panagia points out, we are inclined to "overlook the sensorality of claim making and, especially, the aurality of the utterance" when we focus unduly on language's afterlife as an ephemeral, textual, cognitive system of signs, abstractions, or beliefs.[73] I would add that we should not overlook the sensorality of texts: the texture of their pages and the colors of their inks, and the physical contact between fingers and pens or keyboards through which they are produced. Emphasizing the corporeality of language prevents us from assuming that speech and writing are any less material than the things spoken for. The aesthetic turn moves our thinking about language and politics closer to the rhetorical tradition and its concerns with understanding and promoting the embodied enactment of language in concrete, public fora. As Ernesto Grassi, one of the foremost scholars of the rhetorical tradition (and of Vico), puts it, "rhetoric" is speech that "makes a fundamental contribution to the structure of our world" by "combining in a new way not only words but things as well . . . hence making them evident to us."[74] This is the tradition to which Arendt, Vico, Descartes, and Hobbes belong, or so I shall maintain.

The aesthetic turn and the rhetorical tradition together encourage us to affirm, at once, the embodiment of speech and the creativity of embodiment, and thus to recognize the contributions of speech to the creation of the material world to which our bodies belong. As Zerilli puts it, channeling Arendt, "Rhetoric understood as a quotidian practice of public speech is . . . the condition of our freedom; it opens up the world to us in new ways."[75] Speech creates occasions for transformative politics by making it possible to see and hear not only the speaker, but the world, anew, establishing relationships where none were given before and severing others long established. Such a view of speech's creative potential differs from familiar accounts of the "linguistic construction of reality" because it does not presuppose an ontological disconnect between the immaterial reality that language ostensibly constructs, on the one hand, and the material reality that these linguistic constructions purport to represent, on the other. In different ways, Zerilli, Grassi, and Panagia—and, I argue, Arendt, Vico, Descartes, and Hobbes—all resist the view that speech sweeps us away from the objects and events of the *really* real world and deposits us in a windy Neverland of language. Instead, they suggest that speech intervenes in and adds to the world, constituting new, sometimes unprecedented, worldly configurations.

Perhaps the most radical departure from the language/matter opposition to be found in the aesthetic turn is Rancière's own notion of *le partage du sensible*—the "partition of the perceptible" or "distribution of the sensible." Rancière thinks about membership in a community as a pattern or distribution of visibility and audibility: Who may speak and be heard as a subject? What may appear and be spoken of as a common object? Speech is what maintains (and sometimes disrupts) this pattern, disclosing (and sometimes transforming) the common world. The enactment of speech "defines what is visible or not in a common space."[76] For Rancière, it is by virtue of the fact that some parts of the community have a voice (while others are deprived of any part in speech and debate) that some objects are perceptible and their place in the common world is secure (while others fall below the threshold of perceptibility and have no part in the common world). Rancière's position differs from the conventional constructivist point that language shapes *how* we see things. Rather, the distribution of the sensible delimits *what* we see—"what is visible" or perceptible in the first place, which may, in the second place, be seen and interpreted in various ways by various spectators. Speech contributes to the constitution of the extrasubjective world—not just a subjective or intersubjective worldview—by granting a worldly place to things that would otherwise be imperceptible. Much as Arendt considers the "realness" of the world to be a contingent political outcome, Rancière posits that the very perceptibility of material beings, their susceptibility to being perceived by the five human senses, is constituted through politics, in tumults that make an issue, at once, of the shape of the "we" and the shape of the world. Like Grassi, Rancière sees that speech doesn't just represent reality, but presents it, moving some of its elements to the center of the common world and pushing others to the very margins of perception.

By reorienting us to speech as a sensuous, embodied practice and restoring our appreciation of the body's practical contributions to creating the world to which it belongs, existing works within the aesthetic turn point toward—but stop short of pursuing—a novel approach to the politics of science. Extending their insights, this book attends to the ways that movements to found new sciences or retrench old ones contribute to and compromise the very reality of the world by reorganizing who may speak for the things that appear within it. Natural science is political because it is a particularly consequential way of organizing collective, affective encounters among human and nonhuman bodies. Laboratories, for example, are rich sites for establishing the perceptibility of nonhuman beings and for renegotiating the distribution of voice, transforming who has a legitimate part to play in presenting common objects. Science is both political and aesthetic, then, because it is a potent instance of the embodied activities that stir shared passions and assemble heterogeneous collectives. When its practitioners speak for nature, they both orchestrate and depend on demotic activities and comportments that compose, disclose, and destroy the common world.

Outline of the Argument

My argument begins in earnest in Chapter 1, where I recover a novel critical perspective on the origins, promise, and limits of the authority of science to speak for nature from Arendt's genealogical account of the scientific beginnings of the political problem she calls "earth alienation." To better understand what Arendt might have in mind when she suggests that alienation from earth poses a grave problem for politics, I first reconsider the meaning of "earth" in her worldly vision of politics. Most readers assume that Arendt defines the earth as the source of natural sustenance and the condition for human life, as opposed to "the world," which she defines as the product of human work and action and the condition for life in common. But a closer look reveals that Arendt affirms earth's significance not only as a source of life, but also as a source of diverse sights, sounds, smells, tastes, and textures. Earth qua appearance, I argue, plays an integral role in Arendtian politics; in their givenness, natural phenomena attract the attention of plural spectators, confront them with the specter of otherness, and call upon them to speak and act to constitute a common world. When she accuses Galileo, Descartes, and other early scientific figures of estranging us from earth, then, she has in mind their contributions to reshaping the material circumstances and practices that determine which natural givens appear before whom—the contextual conditions for spectatorship that Arendt calls "common sense." From the beginning, she maintains, the modern natural sciences founded their authority on the refusal to receive earth as it gives itself in appearances and hostility toward common sense and the common world. Earth alienation and world alienation are born side by side, she warns.

Arendt's bleak analysis helps to reveal the relationship between natural science and common sense as a key site of political contest. But because her genealogy treats the assault on common sense as both originative and definitive of the sciences, it tends to occlude the possibility of enlisting and reconstituting scientific authority in the service of sustaining or rebuilding the common world. In Chapter 2, I recover resources to extend and sharpen this Arendtian perspective from another, earlier genealogist of science and politics: Giambattista Vico. Vico anticipates some of Arendt's key insights into the worldly political importance of natural scientists' practices of speech, observation, and experimentation, and he shares many of her concerns that his near contemporaries, especially Descartes, have shored up their authority by repudiating common sense and devaluing the world of politics. But Vico's genealogy casts these worrisome instances of antagonism between natural science and world-building politics as contestable betrayals of a much longer historical legacy of deep and abiding affinities between the act of making the world and the act of knowing it.

I recover from Vico a redescription of common sense as a collective, transgenerational power to create, over time, a series of second natures for human and nonhuman beings alike. From the perspective that I draw out of Vico's writings, all of nature is a historical phenomenon, as much a part of the man-made world as any other political institution. Furthermore, in a novel twist on the maker's knowledge tradition, Vico suggests that the creativity of common sense is the very condition of possibility for the perceptibility of nature to human spectators and the intelligibility of what scientists have to say about the results of their observations and experiments. Through the lens of Vico's genealogy, it is possible to delineate more precisely the complex affinities and antagonisms between various incarnations of natural science and the political worlds to which they belong. Indeed, the combined genealogical guidance of Arendt and Vico makes it possible to recognize political complexities that even they overlooked in their respective accounts of key moments of crisis and renewal of authority in the history of science. In Chapters 3 and 4 I revisit two such moments that Arendt and Vico each consider particularly detrimental to the fate of world-building politics in modernity: namely, the respective efforts of René Descartes and Thomas Hobbes to revolutionize the sciences of nature and of politics.

Arendt and Vico are certainly not the only political thinkers in the near or distant past to have raised alarms about Descartes's hostility toward politics: by virtue of his celebration of introspection, abstraction, and solitude in the *Discourse on Method*, Descartes has become a ubiquitous emblem of the risks of the modern scientific mind-set to democracy and pluralism in contemporary political theory and in the two-sciences literature. But a closer look at Descartes's writings shows that he worked both with and against the political activities, habits, affects, and powers of the people in his efforts to usurp the established authority of Scholasticism and refound the authority to speak for nature on new grounds. My reading in Chapter 3 offers a more nuanced portrait of Descartes as both a solipsist and a publicist; in the *Discourse* and other writings, I show, he made a spectacle of his private acts of thinking by narrating them in the demotic genre of the fable. In the name of reconstituting natural science, Descartes called upon men of common sense to judge his science—indeed, he called upon the people to establish the worldly conditions for their own exclusion from the practices and privileges of science. I recast Descartes's writings about science as acts of politics in Rancière's sense: Descartes audaciously entrusts the task of deciding who should speak for nature to a segment of society that lacks any worldly standing to pass such judgments. But Descartes both opens and closes politics in this sense. He authorizes the people to judge scientific method in order to establish a "partition of the perceptible" in which they would willingly abstain from judging scientific *objects*, the phenomena of nature. I repurpose Descartes as an emblem of modern science's ambivalence between aversion and attraction to common sense, and an illustration of how that wavering can be used to both secure and transform scientific authority. In so doing, I resist the temptation to erase either the democratic or the despotic possibilities of this—or any—example of natural science.

Chapter 4 turns to Hobbes in order to reconsider some of the concerns about the risks of modeling the study of politics on the natural sciences that have encouraged many to embrace versions of the two-sciences settlement. Hobbes's efforts to apply the same geometrical methods to the study of both nature and politics are often recalled as a cautionary tale of the dangers of "scientizing" the study of politics. But Hobbes's critics seldom take the trouble of scrutinizing his geometrical treatises in order to understand what geometry meant to him. Hobbes recognizes, I argue, that a geometrical demonstration can only achieve what I call its "science effect"—its power to compel belief—if it is informed by and rooted in quotidian, corporeal meaning-making practices carried on beyond the confines of the geometer's study. The chapter uncovers a worldly politics of repeated, mutually acknowledged gestures and rhetorical performances at work in Hobbesian geometry. With Hobbes's own unusual view of geometry as a point of reference, it is easier to notice his reliance upon this micropolitics of world-building and meaning-making in his efforts to constitute newly absolute authority for political scientists and political sovereigns. Hobbes enlists the people to "scientize" politics, depending on their world-building practices to manufacture the conditions under which the sovereign's claims about justice would acquire the "science effect" of a compulsory truth. Hobbes's geometrical politics requires that the people remain actively involved in creating and maintaining the worldly conditions for their own subordination. The chapter captures some of the ways that political science in a Hobbesian vein, like natural science in a Cartesian vein, wavers in a constitutive ambivalence between democratic openness and absolutist closure. The ability of sovereigns and of scientists to speak for the world's various human and nonhuman beings is established and distributed—and may be contested and repartitioned—through this wavering.

Finally, the concluding Epilogue draws attention to the contemporary significance of the core insight elaborated throughout the book: namely, that the authority to represent nonhuman nature depends on the sensuous, embodied activities that add meaning, measure, and relation to the material environment and incorporate natural phenomena into the common world. Although the thinkers through whom I develop this insight wrote in the near and distant past, it is particularly important to heed their insights into science's fragile, contestable dependence upon common sense today, when the specter of anthropogenic climate change has created new questions about who could give voice to earth's unprecedented, man-made disorders. The book ends with a call to action that implicates scientists, citizens, and sovereigns alike in the responsibility to rebuild the common world and leave it fit for generations to come to inhabit and to make their own.

CHAPTER 1

Earth to Arendt

"The world" is arguably the central category of Hannah Arendt's political thought and the locus of her hopes and fears about the fate of politics in modernity.[1] As she defines it, the world consists of the physical assemblage of things that appear between men and women, "in the plural," attracting their mutual attention and concern and eliciting their speech, action, and judgment. Arendt's account of the centrality of worldly things (tables, buildings, and artworks, to name a few of her examples) to politics helped to inspire what I call in this book—with Arendt's influence in mind—the "the worldly turn," the recent movement to re-center matter, corporeality, sensation, and affect in political thought. In this chapter, I show how an Arendtian orientation to the world as the material home for politics enables us to recast both nature and the natural sciences as political phenomena of the first order and moves us a few steps beyond the two-sciences settlement—the view that natural and political science should respect the essential differences between their respective objects of study.

Many of Arendt's readers will likely see the intention to politicize nature and the natural sciences in her name as confused or illegitimate. With good reason, Arendt is well known for using the category of the world to *exclude* nature and its sway over the human body from politics.[2] In *The Human Condition*, for instance, she contrasts the world—the artificial product of human work and the site of human action—with "the earth"—the natural givens that humans labor upon and consume in order to survive. Generations of readers have seen Arendt's labor-work-action triad in "'territorial' terms" (as Patchen Markell aptly puts it), as a way of protecting action against the instrumentalism of work and the necessity of labor.[3] Within this territorial frame, Arendt's earth-world dyad seems to further protect action from labor by assigning them to separate ontological spheres. It is tempting to conclude that Arendt's concepts of earth and world reflect a dualistic view of nature and politics.[4] This impression may be reinforced by Arendt's criticisms of political scientists who emulate the methods of natural science for flattening the differences between nature and politics—criticisms that ostensibly capitulate to the oppositional logic of the two-sciences settlement.[5]

But a closer look at Arendt's critical appraisal of the *natural* sciences complicates this territorial impression of the earth-world relationship and reveals tensions between her vision of politics and the two-sciences settlement. While Arendt charges behaviorist social sciences with conflating nature and politics, her primary concern with the natural sciences is their tendency to *disconnect* us from the reality of nature on earth: "Earth alienation became and has remained the hallmark of modern science."[6] Arendt did not accept the two-sciences settlement's reassuring message that natural science is politically irrelevant or benign so long as its techniques are applied exclusively to the study of nature. On the contrary, as the readings to come will show, Arendt faulted the natural sciences for encouraging estrangement from earthly nature, an earth-loss she considered to endanger our political relationships with one another and thus to the common world. According to the genealogy of the modern decline of politics she provides in the final chapter of *The Human Condition* (hereafter cited parenthetically as *HC*), earth alienation and world alienation began together at the birth of the natural sciences, when Galileo first looked through the telescope. In other words, Arendt's discussion of labor in that text is not the end of her story about the earth's significance for the human condition.[7] In the course of this chapter, I excavate the deeper meaning of "earth" to Arendt in order to clarify why she considers scientific alienation from earthly reality so troubling for politics.

I derive an alternative perspective on the Arendtian earth and its significance for the fate of politics in the age of modern science from her unusual account of common sense and its role in organizing collective sensuous encounters with earth's givens. Arendt seldom hews to the conventional meaning of "common sense" as a repository of, well, conventional meanings, beliefs, or opinions.[8] Instead, she charges this "mysterious 'sixth sense'" with fitting us into the world and granting to appearances an unshakable "realness" or "sheer thereness."[9] In these passages, earth's elements and processes represent more than an opposite and a threat to "the world" and to the political activity that it enables and protects. Rather, Arendt emphasizes ways that the earth *shares* with the world the ontological propensity to appear between us. We not only relate to nature qua sustenance when we "metabolize" natural resources as laborers to satisfy the needs of life (*HC* 98); we also receive nature qua appearance as spectators and speak for it as actors, establishing the conditions for life in common. As I read her, Arendt is concerned not only with the risks of natural necessity overflowing its boundaries and invading the common world, but also with the risks of natural displays of diversity disappearing from public view.

By showing that earth, to Arendt, means more than just the condition for life, my readings reveal that more than "mere" survival is at stake for her in the earth alienation she associates with the natural sciences. Arendt's critique casts the natural sciences as antagonistic toward common sense and the common world insofar as participants seek to destroy or obscure earth's gifts of appearance. In her view, the primary political effect of the instruments, experiments, and mathematical techniques of the natural sciences has been to organize spaces of *disappearance*, where earth's visible surfaces are violated and the range of possible perspectives from which they may be perceived is narrowed. She is concerned not only with the potentially life-altering and life-threatening contributions of the sciences to modern technology and industry—the concerns that occupy the few existing studies of Arendt's critique of science.[10] She is also concerned with the sciences' world-threatening potential to devalue human plurality and erase or deface non-human alterity. Arendt's analysis of earth alienation puts to one side concerns about the utility or validity of the sciences in order to press us to question the sciences' implications for the building of a common world—a question that has largely gone unasked by critical thinkers of science and that has largely been overlooked in the secondary literature on Arendt's political theory.

Coming chapters will show that these implications are more complicated than Arendt herself recognized—that the sciences have a promising, if risky, role to play in receiving earth's appearances and augmenting common sense and the common world. But first, we need to better understand the depth and the importance of the concerns that Arendt brings into view when she reframes the politics of science in worldly terms. This chapter develops an Arendtian perspective on science and politics in two stages: In the first part, I discuss the vision of earth qua appearances that emerges from Arendt's discussion of common sense. The second part builds on this revised understanding of the Arendtian earth to recover and extend her critique of natural scientific earth alienation.

Earth

Nature, Among Other Things

The close relationship that Arendt draws between earth alienation and world alienation in her account of the rise of modern science and the decline of modern politics at the close of *The Human Condition* is puzzling in light of the strong distinctions she draws earlier in that text between the concepts of earth and world. To give one example: "The world itself, in so far as it is common to all of us . . . is not identical with the earth or with nature, as the limited space for the movement of men and the general condition of organic life. It is related, rather, to the human artifact, the fabrication of human hands, as well as to affairs which go on among those who inhabit the man-made world together" (*HC* 52). Such passages underscore the difference between the earth, cast as the source of life, and the world, cast as the product of human work and the object of collective political activity. But beyond the earth's contributions to biological survival, Arendt helps us to see that earth qua appearance plays a crucial role in our political flourishing. By appearing, earth solicits our spectatorship and calls us to supplement its colorful displays of diversity by performing our plurality in speech and action.

Nevertheless, a full understanding of the Arendtian earth-world pairing must begin by taking seriously the dire consequences she sees for both life and life in common when certain important differences between earth and world are disregarded. First, Arendt warns that our very survival is imperiled when we allow work or action to unduly interfere with earth's uniquely life-sustaining concatenation of elements, "a habitat" where terrestrial creatures like us "can move and breathe without effort and without artifice" (*HC* 2). She emphasizes that the material conditions of life are *given* on earth, both in the sense that they are not made and in the sense that they are precious, "a free gift from nowhere (secularly speaking)" (*HC* 2) or (less secularly) a "chain of miracles" on par with the miracle of human freedom.[11] We must receive these gifts of life by cultivating, harvesting, breathing, eating, or otherwise metabolizing natural resources in order to survive. But in so doing, we must take care not to destroy our sanctuary in the universe. Arendt thinks it would be a perilous form of hubris to forget that our lives depend on earthly phenomena that we have no hand in making—soil, air, water, and the cyclical processes of birth, growth, death, and decay. Her key example of such forgetfulness is modern utilitarianism, which wrongly extends *Homo faber*'s emphasis on the usefulness of man-made tools to nature, as though these gifts were here for solely our benefit (*HC* 153–59).

Second and conversely, Arendt warns of dangers to life in common when *animal laborans*' preoccupation with the exigencies of biological life supersedes concern for the durable world of our own making. Tables, houses, books, tools, and other human works constitute a world and enable the enactment of politics insofar as they separate and connect those who make and inherit them: "To live together in the world means essentially that a world of things is between those who have it in common, as a table is located between those who sit around it" (*HC* 52). The world is what we have in common even as we contest its meaning or vie for its future. Thus it is the material situation and provocation for political speech and action and their aftermath in storytelling, judgment, and memory: "At the center of politics lies concern for the world. . . . Wherever people come together, the world thrusts itself between them, and it is in this in-between space that all human affairs are conducted."[12] Only in the midst of this in-between does "the human condition of plurality," the "*condition per quam*" of politics, come to fruition, as men and women present themselves in their utter uniqueness, act in concert, and interpret, debate, and narrate what has transpired (*HC* 7). Arendt warns that we can fail to realize the contingent "fact" of plurality and destroy the prospects for politics if we let earthly concerns with the body's needs supersede care for the common world. Such failure has been widespread in modernity, in her analysis, resulting in the displacement of politics by society, "the form in which the fact of mutual dependence for the sake of life and nothing else assumes public significance" (*HC* 46). The rise of mass society represents the "unnatural growth, so to speak, of the natural," as bodily urges and exigencies overtake the world and stifle political activity (*HC* 47).

Thus far, Arendt's world-earth dyad aligns the two worldly activities of work and action while underscoring the importance of protecting both of them against the earthly activity of labor, adding credence to Markell's recent amendment to territorial readings of her labor-work-action triad. As he points out, although Arendt is careful to distinguish the isolation and instrumentalism of work from the plurality and freedom of action, she also recognizes that work "is never *merely* instrumental, but implicates larger questions about, as Arendt would later say, 'how [the world] is to look' and 'what kind of things are to appear in it.'"[13] In a key passage for Arendt's argument about the worldliness of work and Markell's interpretation, she raises the example of the useless work of art to show that all works, apart from their (in)utility, are aesthetically valuable: "Everything that is must appear, and nothing can appear without a shape of its own; hence there is in fact no thing that does not in some way transcend its functional use, and its transcendence, its beauty or ugliness, is identical with appearing publicly and being seen" (*HC* 172–73). Markell rightly argues that, for Arendt, work, like action, adds to the appearing in-between toward which human affairs are conducted, enhancing plurality and politics. Qua appearance, the thing, be it a useless sculpture or a useful chair, "gathers us together" into a thickly tangible public realm and calls to us to appear in word and deed (*HC* 52).

But this key passage also hints at the possibility of an affinity between earth and world—and a deeper challenge to territorial readings of Arendt than even Markell entertains. Trees and rivers, as much as sculptures and chairs, "must appear." As appearances, it would seem, natural elements and processes implicate larger political questions concerning how the worldly in-between is to look. Insofar as earth's givens, like man-made artifacts, transcend their usefulness in appearance, what could their aesthetic and political significance be? In what respects might earth's natural materials and cycles contribute not merely to the survival of human and nonhuman animals, but also to the tangible texture of the common world?

In *The Life of the Mind* (hereafter cited parenthetically as *LM*), Arendt puts the issue of nature qua appearance and its worldly, aesthetic status squarely on the agenda from the very first sentence of the first chapter of the first volume: "The world men are born into contains many things, natural and artificial, living and dead, transient and sempiternal, all of which have in common that they *appear* and hence are meant to be seen, heard, touched, tasted, and smelled, to be perceived by sentient creatures endowed with the appropriate sense organs" (19, emphasis in the original unless otherwise noted). Arendt writes here of a single world, containing both the natural, organic entities and processes that she elsewhere associates with the earth, and the durable, artificial things that owe their existence to human work. Natural and artificial things are united by their shared propensity to appear and thereby to elicit the attention of sentient creatures, including, though not exclusively, human beings. Arendt's immediate concern in describing this single, appearing world is to heal the traditional metaphysical divide between "(true)

Being and (mere) Appearance" (*LM* 23); in a world where "*Being and Appearing coincide*," philosophers' habit of seeking true being above or behind mere appearance is misguided (*LM* 19). But this marriage between being and appearing also establishes an intimacy between (natural) earth and (artificial) world: Arendt posits that everything that is, given and made, also appears. Earth gives us not only the gift of life, but also the gifts of sights, smells, sounds, textures, and tastes; it calls us to receive its givens not only by metabolizing them in labor, but also by seeing, smelling, and otherwise perceiving them. Natural phenomena are meant to be consumed *and* "meant . . . to be perceived," and this second, aesthetic purpose binds earth and world together in a shared fate.

Attending to earth's aesthetic dimensions helps us to recognize that, as sentient creatures, our duties toward earth go beyond *animal laborans*'s self-interested concern to avoid damaging earth's life-sustaining processes. While we depend on earth qua sustenance to survive, earth qua appearance depends on *us* to perceive and thereby to world it: "Dead matter, natural and artificial, changing and unchanging, depends in its being, that is, in its appearingness, on the presence of living creatures. Nothing and nobody exists in this world whose very being does not presuppose a *spectator*. . . . [S]entient beings—men and animals, to whom things appear . . . as recipients guarantee their reality" (*LM* 19). Earth's creatures are responsible for securing the very reality of the natural and artificial things they perceive; our spectatorship ensures the coincidence between being and appearing that is, for Arendt, characteristic of the world. This responsibility is inscribed in our bodies, in the "astounding diverseness of sense organs" given—as if a gift from nowhere—to earth's diverse species, equipping us to receive and thereby to world "the almost infinite diversity of its appearances, the sheer entertainment value of its views, sounds, and smells" (*LM* 20). Just as all appearing beings are meant to be perceived, all sentient beings are meant to perceive them. Arendt suggests that sentient life, human and otherwise, is always already implicated in worldly purposes that exceed, even as they presuppose, individual survival. The establishment of the world requires that the diverse creatures of the earth enjoy both its nutritive, survival value and its aesthetic, "entertainment" value.

The earth, Arendt reminds us, is the only planet known to be populated with "creatures able to acknowledge, recognize, and react to—in flight or desire, approval or disapproval, blame or praise—what is not merely there but appears to them" (*LM* 19). In this respect, the earth is a uniquely hospitable environment for the world, providing for both astoundingly diverse sentient life-forms and almost infinitely diverse spectacles. On a lifeless planet, other kinds of beings might well appear, but they would lack the fullness of being of the world sustained on earth. The two purposes that Arendt ascribes to all sentient creatures, to live and to perceive, are thus mutually entwined, for the struggle to live—including the toil of human labor—is always already a struggle to maintain the world's fragile coincidence between appearing and being. We might call these struggles "the *natural* growth of the natural": where Arendt warns that it is "unnatural" and dangerous to give undue priority to issues of natural necessity in politics, she also affirms our role as sentient creatures in guaranteeing the reality of nature's appearances.

"Common sense" is Arendt's name for our capacity to fulfill this responsibility to world the earth: "It is by virtue of common sense that the other sense perceptions are known to disclose reality and are not merely felt as irritations of our nerves or resistance sensations of our bodies" (*HC* 208–9). Like eyes, ears, and other sense organs, "common sense . . . and the feeling of realness belong to our biological apparatus" (*LM* 52). She views this sixth sense, like the other five, in physical terms as an embodied power to receive sensuous evidence of the existence of appearing things—unlike *sensus communis*, which, as Arendt appropriates it from Kant, belongs to the mind.[14] Where mental reflection "deals with invisibles, with things not *present* to the senses" and establishes distance between the thinker and the world (*LM* 51), common sense "fits us into" the world, making it impossible to ignore the sensuous presence or sheer thereness of its appearing beings.[15] Where fingers feel textures or eyes feel colors, common sense feels realness—a feeling that "actually accompanies all the sensations of my senses," but to which they are each, alone, insensate (*LM* 50–51).

But although common sense belongs to the fabric of the human body, it is not "merely" biological or natural and "cannot be localized as a bodily organ" (*LM* 50). Though corporeal, common sense is not *given* by nature—as if a gift from nowhere—as eyes, ears, or other organs are. Rather, common sense must be *established*, instantiated through activities that organize the multiple sense organs of multiple sentient beings to receive the same appearances in tandem: "The 'sensation' of reality, of sheer thereness, relates to the *context* in which single objects appear as well as to the context in which we ourselves as appearances exist among other appearing creatures" (*LM* 51). On Arendt's account, common sense satisfies the contextual prerequisites for realness by forging commonalities across two layers of difference: (1) the individual spectator's disparate sense organs, and (2) the disparate spectators who gather together around the same appearing thing.[16] In this way, the establishment of common sense ensures (1) that "it is the same object that I see, touch, taste, smell, and hear" and (2) that I am not the only one to see, touch, taste, smell, or hear it (*LM* 51). The act of gathering together so that things appear *between* us is what impresses their reality upon each of us: "Our 'perceptual faith,' as Merleau-Ponty has called it, our certainty that what we perceive has an existence independent of the act of perceiving, depends entirely on the object's also appearing as such to others and being acknowledged by them" (*LM* 46). In other words, reality would be just as elusive on a planet populated with only one sentient creature (or with creatures who scorned one another's company) as on a planet populated with none. Common sense makes use of the diversity of appearances and spectators given on earth to create the contextual conditions for the reality of the common world to be secured—the natural growth of the natural. Arendt's account of common sense suggests that the capacity of nature's appearances to relate and separate and the capacity of sentient creatures to organize contexts of simultaneous spectatorship are mutually dependent and reinforcing.

Fortunately, on earth there is no shortage of sentience.[17] Common sense brings the natural fecundity and variety of earth to fruition by drawing some of its many life-forms into shared spectatorial contexts. Indeed, Arendt treats the very diversity of species on earth as a third opportunity for common sense to establish commonality across difference: "In a world of appearances, filled with error and semblance, reality is guaranteed by this three-fold commonness: the five senses, utterly different from each other, have the same object in common; members of the same species have the context in common . . . and all other sense-endowed beings, through perceiving this object from utterly different perspectives, agree on its identity" (*LM* 50). Our sense of the realness of the *human* world is enhanced by our sensuous confrontations with the fact that human perspectives, diverse though they may be, are far from exhaustive. Indeed, earth's species are so "utterly different," endowed with such an incommensurable array of sense organs, that common sense strains against its own limitations in striving to draw these species closer together and fit them into a common world: "Every animal species lives in a world of its own" (*LM* 20). But although Arendt considers us to be outsiders to the worlds of bats or lizards, with their utterly different organs of echolocation or infrared vision, she also maintains that this outsider status is itself an asset to common sense and the realness of the human world (and, presumably, the worlds of bats and lizards). The otherness that divides sentient creatures into separate species-worlds is among earth's most valuable gifts, challenging members of each species to perceive identity across even radical difference.

Earth's diversity of species is also an asset to common sense because it gives us so many entertaining sights and sounds to attract the shared attention of more than one creature at a time. Arendt, inspired by zoologist Adolf Portmann, is especially impressed by "the enormous variety of animal and plant life, the very richness of display in its sheer functional *superfluity*"—for example, "the plumage of birds . . . 'the intrinsic worth of which lies solely in its visible appearance.'"[18] For Arendt and Portmann, birds are colorful because, like humans, they are meant to see and be seen, purposes that only seem superfluous within a narrowly functionalist framework of survival or utility. Within Arendt and Portmann's alternative, aesthetic frame, it becomes possible to recognize the *intrinsic* worth of displays by birds, flowers, humans, and other life-forms. These spectacles are valuable because they call us to take up our role as spectators, exercise common sense, and thereby contribute to the realness of our various species-worlds. Life itself is not "mere" or bare life to Arendt; rather, "On this level of sheer being alive, appearance and disappearance, as they follow upon each other, are the primordial events" (*LM* 20). And earth is not "merely" the material cause of these primordial aesthetic events, life and death, but also the stage on which they are performed and enjoyed: "To be alive means to be possessed by an urge toward self-display which answers the fact of one's own appearingness. Living things *make their appearance* like actors on a stage set for them" (*LM* 21). This stage is set, first, by nature, which organizes matter into appearing bodies, some of which are vested with organs and urges to see and be seen. Common sense continues what nature begins, organizing diverse living bodies and their diverse sense organs to constitute shared scenes of spectatorship. This "mysterious" sixth sense transfigures the earthly stage into the real world: the natural growth of the natural.

Work, Action, and "the Law of the Earth"

Without conflating "earth" with "world," then, the opening pages of *The Life of the Mind* and Arendt's comments on common sense throughout her corpus complicate the comparisons between these categories drawn in *The Human Condition*. Earth qua appearance gives us meaning and purpose, enjoining us, like all sentient creatures, to see and be seen and enabling us to guarantee the reality of the world. This richer understanding of the growth of the world from the earth equips us to complicate, too, territorial readings of work and action as activities that disconnect us from earth and transport us to an artificial world. Work and action, as we shall see, surpass earth's givens by extending or augmenting them, and it is precisely earth, in its dazzling aesthetic diversity, that inspires this surpassing.

In her enigmatic final sentence of the first paragraph of *The Life of the Mind*, Arendt suggests that the earth impels the human activities that make politics possible: "Plurality is the law of the earth" (19). What does it mean to say that plurality, the defining quality of the world and the sine qua non of politics, is the *law* of the earth? Plurality, as we have seen, is *not* a natural given for Arendt, but a contingent fact of the human condition—a fact that is realized through work and action and that can be eroded or destroyed when these human activities are devalued or abandoned. However, we have also seen that *diversity* is a defining attribute of the Arendtian earth, with the "almost infinite" array of sights, sounds, smells, textures, and tastes it offers to creatures born hungry for sensation and eager to display themselves. By describing plurality as "the law of the earth," Arendt casts the aesthetic diversity of earth qua appearance as a normative, law-like injunction requiring human actors to constitute the artificial conditions for politics. She suggests that the purposes we share with other sentient creatures—to see and be seen—also point us toward a related, but distinctively human, ideal: plurality, the revelation in word and deed of the utterly unique "who" that each human subject is. Worldly politics, on this reading, is neither opposite nor identical to earthly nature. Rather, the earth gives the law to the world. The Arendtian law of the earth urges human creatures to become creators—to surpass, even as they imitate, nature's "richness of display" by acting to enable more things to appear from more perspectives.

Read in this light, the activity of work promises to help build upon the diversity given by nature by establishing the artificial conditions for things to appear for a long time. As Arendt argues, the longer a thing lasts, the better its chances of attracting an audience and fulfilling its fragile potential to belong to the world: "Reality in a world of appearances is first of all characterized by 'standing still and remaining' the same long enough to become an *object* for acknowledgment and recognition by a *subject*" (*LM* 45–46). Work reorganizes earthly matter into the artificially durable form of an object, ensuring that at least some things will appear much the same tomorrow as they do today and did in the past. Reading Arendt's comments on work and on common sense in tandem suggests that work artificially supplements the human sense of realness by constituting, beyond the three layers of diversity given by nature, a fourth, intergenerational layer. Thanks to work, the reality of what I and my contemporaries perceive between us is further confirmed when we see that it has been there long enough to have been perceived by past generations and could well remain for generations to come. The fabrication of durable objects from natural materials expands the context of spectatorship across time, making an in-between out of the gap between past and future. From this Arendtian perspective, work destroys earth's givens in the service of following earth's law, intensifying the realness of whatever appears in the moment.

In so doing, work enhances the realness not only of its products, but also of the natural materials and processes it leaves untouched. The temporal gap opened by work provides a window through which even earth's most ephemeral phenomena may be viewed and from which they, too, derive the special, durable thereness of objects: "Only we who have erected the objectivity of a world of our own from what nature gives us, who have built it into the environment of nature so that we are protected from her, can look upon nature as something 'objective.' Without a world between men and nature, there is eternal movement, but no objectivity" (*HC* 137). Not only the table where I write or the page from which you read, but also the living and dying trees outside of our respective windows appear within the objectivizing freeze-frame of the durable human artifice. Tables, books, buildings, and other works serve as an in-between that relates and separates not only men and women, in the plural, but also "men and nature," in the almost infinite diversity of their appearances. Even living bodies, possessed by nature of "an urge to self-display" but condemned by nature to disappear after a short time, may attain the intensified realness of objectivity when they appear alongside artificial things, which will outlast even the humans who made them (*HC* 137). The products of work are artificial, but not—in Arendt's pejorative sense—unnatural when they help to secure the reality of all things, given and made, and improve their suitability to separate and connect men and women, in the plural.

But it is in action that earth's law, plurality, comes closest to fulfillment. It may be difficult to see how Arendt considers earth's appearances to be inducements to speak and act, given her firm contrast between human action, on the one hand, and the "fact of appearingness" that characterizes all things and bodies, on the other: "Speech and action . . . are the modes in which human beings appear to each other, *not* indeed as physical objects, but qua men" (*HC* 176, my emphasis). She holds that human actors do not merely appear, as all things do, nor do they merely display, as all life-forms do. Moreover, humans who act actively "express," "reveal," or "disclose" themselves, answering the question "Who are you?" (*HC* 178 and passim). Yet she also points out that such self-revelation transpires by magnifying, not by erasing or occluding, "the curious quality of *alteritas* possessed by everything that is" (*HC* 176). This "otherness . . . is an important aspect of plurality," but it only comes to fruition when human speech and action render it explicit and articulate: "Only man can express this distinction and distinguish himself. . . . In man, otherness, which he shares with everything that is, and distinctness, which he shares with everything alive, become uniqueness, and human plurality is the paradoxical plurality of unique beings." Action, for Arendt, at once surpasses and extends the diversity given on earth, converting it into plurality.

Conversely, earth inspires and impels us to act through its profligate gifts of sights, sounds, smells, tastes, and textures. This connection between earthly life and political action is marked by Arendt's famous metaphor likening action to "a second birth, in which we confirm and take upon ourselves the naked fact of our original physical appearance" (*HC* 176–77). In action, we respond to our bodies' givenness by confirming and seizing hold of it, not by escaping or negating it; we restage the primordial aesthetic event of our own births. The law of the earth demands that human actors do what no other being or creature on earth can do: to perform and give voice to the alterity that would otherwise appear, in the physical, "naked fact" of human and nonhuman bodies, but remain inarticulate, unspoken. For Arendt, plurality is realized when human action reveals the distinctions that both tie human beings to and set them apart from every other "what" and "who" on earth.

When we fulfill the law of the earth and appear "qua men" in action, we contribute to the constitution of a "space of appearances," a public realm where men and women gather to reveal themselves and to witness one another's words and deeds. This is the space of politics that Arendt seeks to protect against "the unnatural growth of the natural," fearing that undue attention to bodily necessity will erode care for the common world. But the space of appearances is both a refuge from earth qua sustenance and an extension of earth qua appearances. As we have seen, earth, to Arendt, is a stage—in this regard, the earth is a primordial principle and inspiration for the theatrical space of politics, where men and women perform their uniqueness before an audience of equals. Action augments common sense by organizing spaces of heightened visibility, where both appearances and perspectives proliferate. It is as though action fits the organless common sense with glasses, enabling spectators to percieve the realness not only of physical things, but also of "intangible" words and deeds (*HC* 183): "The physical, worldly in-between along with its interests is overlaid and, as it were, overgrown with an altogether different in-between which consists of deeds and words and owes its origin exclusively to men's acting and speaking directly *to* one another. . . . We call this reality the 'web' of human relationships" (*HC* 182–83). This passage is often read as though it committed Arendt to an ontological division between two kinds of reality: one physical, the other performed. But her distinction between the performative and the physical layers and entwines them. Arendt cannily likens their intimacy to a plant growing from and rooted in the earth's surface, and she affirms this growth of the natural into the political. As Arendt depicts it, action adds to the world the web-like reality of human relationships without detracting from the reality of material things.

On the contrary, as Arendt's "web" metaphor suggests, action has the effect of sticking human subjects all the more firmly to the reality of the physical objects, natural and artificial, in their midst. Those who act reveal at once who they are and what in the world they care about, their interests or "*inter-est*, which lies between people and therefore can relate and bind them together. . . . Most words and deeds are about some worldly objective reality in addition to being a disclosure of the acting and speaking agent" (*HC* 182). Our acts of self-disclosure disclose, too, something of the *alteritas* of the things that interest us: "We are unable to say what anything is without distinguishing it from something else" (*HC* 176). Human speech thus rescues the inarticulate otherness of nature from obscurity by giving it voice. In presenting and speaking for ourselves, we also present and speak for other, inarticulate and voiceless beings. So even as action introduces a new, intangible reality into the world, it also enhances the reality of physical, tangible objects, both given and made: "Without a space of appearance and without trusting in action and speech as a mode of being together, neither the reality of one's self, of one's own identity, nor the reality of the surrounding world can be established beyond doubt" (*HC* 208). Even natural phenomena are more securely real when they interest human actors than in any other context. The law of the earth calls us to give voice to the natural diversity that tangibly appears, but remains unspoken, as it is given.

This law-like call to speak for ourselves and for the things in our midst is itself unspoken; the earth can only convey its demand that we act to realize plurality by appearing, in all its diversity. Earthly nature calls to us not only to satisfy the private urges of the body, but also to step into public and to reveal and witness our own and each other's relationships to earth's profligate sights, sounds, smells, textures, and tastes. Reading *The Human Condition* and *The Life of the Mind* in tandem, we see that the human world is not a reality apart from the earth, as though only man-made things could be real to us while natural things are real for other species. On the contrary, as I read her, Arendt tells us that common sense fits human beings into a heterogeneous world, a reality only partially of our making. At the same time, she suggests that humans are the only animals to artificially enhance their common sense through work and action, thereby intensifying the reality of all things, given or made. Thanks to common sense, the earth's beautiful and strange appearances need not be any more alien from the human world than they are from the worlds of bats, snakes, or birds. Indeed, natural phenomena may be more real within the human world than within the world of any other species, because humans are the only sentient creatures on earth with the ability to act to enhance common sense.

But, she warns, the human species is also uniquely capable of devising ways to numb common sense and become deaf to the call of the law of the earth. Next, I turn to Arendt's critique of the development that represents, in her view, the greatest threat to common sense's grasp on earthly reality and a serious danger to the fate of politics in the modern age: the rise of the natural sciences.

Earth Alienation

Arendt is all but ignored in science and technology studies, and her ideas about science are seldom a topic of much interest among her readers in political theory. These oversights are surprising given the prominent place Arendt accords to the theme of science.[19] For example, she opens *The Human Condition* by evoking *Sputnik*, the splitting of the atom, and the "attempt to create life in the test tube" as cautionary examples of the "desire to escape from imprisonment to the earth" and rebel "against human existence as it has been given," and she closes this text by lamenting that scientists' awesome and terrible powers to transform the face of nature may be the last remaining vestiges of the human capacity for action (1–2, 324). The few studies to attend to these warnings about scientific earth alienation tend to read them in terms of the narrow view of earth qua sustenance that I have been examining.[20] Because these readers see the Arendtian earth primarily as a source of life, they tend to approach the problem of earth's repudiation in terms of the sciences' interventions into biological survival and reproduction. For example, Kimberley Curtis approaches Arendt for help in defending "elemental characteristics of human life" against modification by reproductive technologies.[21] Arendt is indeed concerned that scientists may someday succeed in "cutting the last tie through which even man belongs among the children of nature" (*HC* 2). But such readings tend to elide her more immediate—and more immediately political—concerns about the role of past and present natural scientists in damaging and devaluing common sense, cutting the tie to earth qua appearance. What is ultimately at stake for Arendt in science's repudiation of earth's gifts of sights, sounds, and smells is not just human survival, but human speech and its power to reveal, a "crisis" that is "political by definition, for speech is what makes man a political being" (*HC* 3).

Galilean Glances

In *The Human Condition*, Arendt seeks out the defining features and enduring significance of the sciences genealogically, from their earliest moments of inception. Like many stories of scientific origins, hers leads back to one Galileo Galilei, whose "first tentative glances into the universe" through the telescope Arendt accounts as one of the "three great events" that "stand at the threshold of the modern age and determine its character" (257–58, 248). But Arendt's portrayal of Galileo subverts the characterization of modern science that his story is typically evoked to support—namely, the view of science as the triumph of empirically validated truths over religious superstition and ideology. According to this received view, Galileo's words mattered when he stood before his persecutors at the Inquisition and whispered his courageous (if apocryphal) testimony to the earth's truth—"and yet, it moves!"—because he had validity on his side, having adhered to an early version of the scientific method. Arendt's version of Galileo's tale cautions against this tendency to conflate scientific authority with empirical validity. She retraces Galileo's authority to speak for nature to a different source: his use and abuse of common sense. In her telling, Galileo established new authority by establishing a new, artificial sense for the real: the telescope. But, she warns, the telescope at once aided and undermined common sense, rendering scientific authority and the feeling of realness on which it depends contingent and fragile: "Despair and triumph are inherent in the same event" (*HC* 262). By decentering validity from our understanding of scientific authority, Arendt's story of Galileo equips us to better understand why scientific speech often tragically fails to reveal—or even alienates us from—reality even when scientists are speaking the truth.

To be clear, Arendt does not raise these complications in Galileo's relationship to truth and realness in order to raise doubts about the validity of his representations of nature. On the contrary, she performs her participation in what she considers to be the unshakable contemporary consensus around heliocentrism. For example, in "Truth and Politics," she uses the sentence "'The earth moves around the sun'" to exemplify how some statements, "once perceived as true and pronounced to be so," are "beyond agreement, dispute, opinion, or consent."[22] In this essay, as in *The Human Condition*, she is interested in understanding—not challenging—this authority, interrogating its political sources, limits, and consequences. She takes for granted both that heliocentrism is valid and that this taken-for-granted status cannot be explained by validity alone. In other words, the question of truth's authority is irreducible to epistemic considerations and requires investigation from a political perspective.

So while Arendt implicitly acknowledges, in the context of "Truth and Politics," that one aspect of Galileo's legacy was his role in making it unthinkable to doubt the earth's motions, she also implicitly resists, in the context of *The Human Condition*, the conventional view that Galileo's validation of the truth of the moving earth was the defining event of modern science. For one thing, she is keen to remind us that Galileo was by no means the first to demonstrate or proclaim it to be true that the earth moves: heliocentrism had already been hypothesized by Nicholas of Cusa and Giordano Bruno and mathematically modeled in vivid detail by Kepler and Copernicus on the basis of voluminous astronomical observations. These thinkers "needed no telescope to assert that . . . the earth circles the sun," yet—until Galileo—no such assertion "has ever constituted an event" (*HC* 258–59). Far from questioning the truth of Galileo's own published and spoken testimony in favor of heliocentrism, Arendt proposes that truth was not its most salient feature. Rather, she suggests, in order to understand why Galileo's speech was uniquely impactful and consequential in both the history of ideas and the history of politics, we must reexamine his acts of spectatorship outside of the epistemological frame in which they are typically considered.

In Arendt's alternative framing, Galileo's unique accomplishment was to corporeally perceive what he and others already intellectually believed: namely, that the earth moves around the sun. To pull off this unprecedented feat, the artificial magnification of the telescope was indispensable: "What Galileo did and what nobody had done before was to use the telescope in such a way that the secrets of the universe were delivered to human cognition 'with the certainty of sense-perception'; that is, he put within the grasp of an earth-bound creature and its body-bound senses, what had seemed forever beyond his reach" (*HC* 259–60). One eye closed against the sight of his surroundings, the other wide open at the telescope's aperture, Galileo saw things that had never appeared to any creature on earth. As Arendt uses Galileo's own words to emphasize, his telescope borrowed "'the certainty of sense-perception'" from the very sense organs that it modified (*HC* 260). To put this in terms with which we are by now familiar, the telescope altered and enhanced common sense—the profound, unshakable feeling that the things we sense are really there. Arendt adds Cardinal Bellarmine's words to the discussion to make the point: Galileo was the first "'to demonstrate the reality of the movement of the earth,'" thus earning the ire of an Inquisition unconcerned with mere demonstrations of truth (*HC* 260). By introducing the telescope to his "body-bound" endowment of sense organs, Galileo at once drew upon the commonsense feeling of realness and magnified it, extending the reach of the organless sixth sense beyond the earth to grasp the reality of almost impossibly remote astronomical phenomena.

Returning to *The Life of the Mind* helps to highlight the difference between Galileo's newly felt certainty in the reality of the earth's motions and the intellectual beliefs held by Galileo and his predecessors before anyone ever picked up a telescope. The certainty of the senses, for Arendt, is corporeal and nonepistemological; it is felt by the body and guaranteed by common sense regardless of whether it is ultimately questioned, confirmed, doubted, or believed by the cognitive faculties to which it is delivered. Indeed, it is by suspending all such cognitive questions about validity that common sense "fits" us into the common world: "There we are and no questions asked" (*LM* 59). This is not to portray the body-bound senses and common sense, their mysterious leader, as incapable of seeing through illusory appearances and voiding their guarantees of realness. Rather, Arendt clarifies (quoting

Merleau-Ponty) that such moments of disillusionment are "'always for the profit of a new appearance which takes up again for its own account the ontological function of the first'" (*LM* 26). Galileo's telescopic observations had this ontological function, replacing one appearance with another and establishing the moving earth as a new reality. This ontological displacement occurs alongside of and bears some obvious affinities with the epistemological function of validating the heliocentric hypothesis with new evidence. But Arendt also wants us to see that the ontological and the epistemological aspects of science are not reducible to one another. The moving earth as it is imagined, contemplated, and calculated by astronomers, past and present, is an unreal thought-thing—the product of "a mental act," which "always transcends the sheer givenness of whatever may have aroused its attention" (*LM* 74). Galileo was the first to extend, not transcend, the power of the "body-bound" senses, which is why he alone was able to rescue the moving earth from the unreality of the "realm of ideas" (*HC* 260). By virtue of the telescope, the earth's movements were not "merely" true, but also real, as tangible in their sheer givenness as anything else disclosed to the senses. And Galileo's words describing these movements took on the weight of reality, lending them the compulsory force to contest geocentric dogma. These words derived their "science effect," as I call it, from the power of the senses to receive and guarantee reality.

To some extent, Arendt believes, the sciences retain this original Galilean affinity between truth and realness as a kind of birthmark: "Science's basic goal—to see and to know the world as it is given to the senses—and its concept of truth is derived from the common sense experience of irrefutable evidence, which dispels error and illusion" (*LM* 58). Before it veers toward tragedy, her story of Galileo points toward the possibility that the sciences—like other ways of organizing human work, words, and deeds—could play a potent and triumphant role in augmenting common sense and fulfilling the law of the earth, its call to pluralize who and what may appear and to whom. Indeed, her portrait of Galileo's realignment of truth and realness at the origins of modern science invites comparison with her celebration of the world-building accomplishments of the original truth-teller, the first historian, Herodotus. "No human world destined to outlast the short life span of mortals within it will ever be able to survive without men willing to do what Herodotus was the first to undertake consciously—namely . . . to say what is. No permanence, no perseverance in existence, can even be conceived of without men willing to testify to what is and appears to them because it is."[23] Herodotus founded ancient history by acting as a spokesperson for the reality of that which appears, affirming with his truth-telling the fragile marriage between being and appearing that, for Arendt, constitutes the world and sustains life in common. She eulogizes Herodotus for giving voice to the irreducible *alteritas* of events that would otherwise have disappeared into the darkness of the past, and thereby contributing stability and permanence to the human world. Although historians in the Herodotean tradition—in contrast with the political actors whose deeds and words they record—seek to put aside their particular interests in or opinions about the things they perceive, they do not put aside appearances as such, on Arendt's analysis. On the contrary, Herodotean truth-tellers seek to minimize (even if they can never completely eliminate) the traces of their particular identities or perspectives in their speech in order to more fully devote their voices to amplifying the sheer thereness of things that appear in different guises to different spectators from different perspectives.

Reading in tandem Arendt's respective assessments of the origins of modern science and ancient history helps us to see Galileo, like Herodotus, as a figure who used his platform as a truth-teller to rescue appearances from oblivion, testifying to the sheer thereness of phenomena so remote as to be invisible to the naked eye. His words both borrowed the force of reality from the things he had seen and returned to them the greater visibility of an object of public scrutiny. Latent within Arendt's critical appraisal of modern science is a glimmer of hope that Galileo's introduction of a novel instrument of magnification into an antique tradition of frank speech could contribute astronomically to a world-building politics.

But Arendt ultimately holds Galileo responsible for dashing these hopes. Galileo was both the first to feel the reality of the earth's motions and the first to experience the dizzying despair of believing that everything that feels real is false: "It is as if Galileo's discovery proved in demonstrable fact that both the worst fear and the most presumptuous hope of human speculation, the ancient fear that our senses, our very organs for the reception of reality, might betray us, and the Archimedean wish for a point outside the earth from which to unhinge the world, could only come true together, as though the wish would be granted only provided that we lost reality and the fear was to be consummated only if compensated by the acquisition of supramundane powers" (*HC* 262). The prospect that our senses betray us is inherent in their artificial augmentation, Arendt warns. As soon as he put down the telescope, Galileo returned to his unaided five senses and felt, in its unmistakable givenness, the terrestrial reality belied by his instrument: the sun moving slowly overhead and the ground unmoving beneath his feet. Of course, Arendt recognizes that cognitive doubt about the reliability of the senses is at least as old as ancient philosophy. But Galileo was the first to feel with "the certainty of sense-perception" just how *uncertain* even the most unambiguous testimony of all five senses—and yet it is still!—can be. One cost of grasping the reality of the universe was a loosened grip upon the mundane reality given to our senses here on earth. Quite unlike the reality-stabilizing effects of Herodotean history, Galilean science set the earth spinning, inducing the vertiginous, stomach-churning feeling that the human sense organs do not fit us for life on this planet after all.

This radically disorienting conflict between what we feel to be real and what we believe to be true is what Arendt has in mind when she speaks of "earth alienation." Earth alienation names disillusionment of a higher order than the familiar feeling that my eyes or ears are tricking me as they compete to receive the superfluity of semblances given by earth qua appearance. The five body-bound senses regularly report each other for small acts of betrayal. But the familiar feeling of befuddlement produced by these reports seldom scares us dizzy, though it may be frustrating (or funny). *Common* sense, as we have seen, ordinarily steps in to convert this mundane sibling rivalry among the senses into the feeling of realness, the unshakable sensation that the things we smell, taste, see, hear, and feel, however variously and imperfectly, are really there. But Galileo's introduction of the telescope assailed not merely the reliability of this or that sensation, but the hitherto "unquestioned assumption that the senses as a whole—kept together and ruled over by common sense, the sixth and the highest sense—fit man into the reality which surrounds him" (*HC* 274). The telescope, Arendt argues, impugned not just this or that sense organ, but the very guarantees of common sense that these organs, despite their fallibility, tether us to something real. The observational practices of modern science, beginning with Galileo, are at once parasitic upon and destructive of "the certainty of sense-perception," she warns. Earth alienation is a symptom of the injury inflicted on common sense by the artificial "sixth sense" of the telescope.

Galileo's telescope dealt more than one blow to common sense, striking at once at the coordination of the five organ-based senses and the organization of various spectators into a shared context. As we have seen, for Arendt, the feeling of realness depends on each spectator perceiving that she is not the only one to see, hear, or otherwise sense appearing things. But the telescope at once enhanced and blinkered Galileo's vision, temporarily disappearing his visible surroundings—including the presence of other people around him—even as it magnified the invisible reaches of space. Telescopes promote tunnel vision; though Arendt herself does not argue this point, she perhaps has it in mind when she uses the plural to lament that "*we* lost reality" with Galileo's glance skyward (*HC* 262, my emphasis). The telescope eroded our collective assurances that the same reality appears to each of us, however differently we perceive it from our various perspectives. It displaced the diversity and debate of a plural context with the Archimedean standpoint, a singular "point outside the earth" (*HC* 262). Read in this way, Arendt's genealogy makes Galileo's telescope available as a metaphor for the narrowing of the field of vision (and hearing, smell, etc.) that modern science accepts as the price of attaining an approximation of the decontextualized view-from-nowhere. We can begin to see why world alienation and earth alienation would be born as twins in this genealogy: intimacy with the invisible phenomena of the universe cannot engender the worldly relationships among spectators fostered by earth's appearances because these astronomical phenomena can only appear to one (instrumentally enhanced) eye at a time.

Finally, in a third blow to common sense, the telescope represents an unprecedented assault upon the distinctively human capacity to augment plurality across time through the activity of work and its durable, artificial products. Arendt tells us that the telescope irreparably damaged the intergenerational connections that durable artifacts otherwise work to secure: it revealed that "the human eye can betray man to the extent that so many generations of men were deceived into believing that the sun turns around the earth" (*HC* 275). After Galileo, even the plurality of generations—the fact that men, not man, have long inhabited the earth, and will for some time—lost its reliability as a guarantor of the reality of appearances. The Galilean event narrated by Arendt thus illustrates, by way of its destruction, that the formerly robust sense of spectatorial solidarity across time was never secured by man-made artifacts alone. Earth's repetitive astronomical cycles of day and night and summer and winter also provided a thread of continuity between past, present, and future generations of spectators, tying them together into a shared context. But the introduction of artificial vision into the human sensorium imperiled both this circadian sense of kinship and the artificial bonds among generations that are literally built through the human activity of work, Arendt suggests.[24] This is another reason that earth-loss gives rise to world-loss: when the very stability of earth's surface is debunked as an optical illusion, then all that is solid, durable, lasting, or reliable—including the man-made infrastructure of the human world—begins to dissolve, swerve, or shake. Neither the human world nor the politics it sustains can survive unscathed by the rise of a science that would repudiate common sense.

Nor can the authority of the scientist who inflicted these injuries. Arendt's genealogy helps to bring to light the peculiar, unenviable position that Galileo found himself in when called before the Inquisition: having opened a gap between truth and realness with his telescope, Galileo was forced to make a Sophie's choice between them. In order to tell the truth, he had to speak out not only against the official dogma of the church, but also against the reality of earth as it gives itself in appearances to every sentient creature under the sun. "The movement of the sun" is what Arendt calls an "authentic semblance": it "will not yield to any amount of scientific information, because that is the way the *appearance* of sun and earth inevitably *seems* to an earth-bound creature" (*LM* 38). Today, thanks in part to the information Galileo gathered with his telescope and provided, at personal risk, to the public, the statement "the earth moves around the sun" binds our beliefs. But although the validity of these words is beyond doubt, their authority is thin: they cannot reach to bind our bodies, which are unyielding in their physical attachment to earth qua appearances. Although it is as unthinkable to me as it is to Arendt to deny the truth of the moving earth, this does not change the fact that I feel down to my bones that I am sitting still as I write this. This feeling of stillness is so unyielding, Arendt points out, that even "the modern scientist" equipped with an arsenal of instruments descended from Galileo's telescope cannot destroy "his own sensation of reality, telling him, as it tells us, that the sun rises in the morning and sets in the evening" (*LM* 53). In this deeper, embodied respect, even Galileo could not have convinced himself of the reality of the moving earth.

Galileo's authority to speak for the things he had seen was thus fractured in the kiln in which it was fired. Although true, his words could not reveal, for the revelatory power of human speech was always premised on receiving and enhancing the appearances of things that show themselves. Galileo's glances through the telescope deprived his words of the power to physically connect men and women, in the plural, to the things that appear between them. This is a crucial part of what Arendt has in mind when she writes in *The Human Condition*'s opening pages that "scientists *qua* scientists . . . move in a world where speech has lost its power" (*HC* 4). Unlike Herodotus, the more faithfully Galileo spoke for the things he had seen with "the certainty of sense-perception" through the telescope, the more his words compromised the fragile marriage between being and appearing that holds together the human world on earth. The original and defining difference between natural science in the Galilean vein and history in the Herodotean vein, as Arendt presents it, is not that scientists and historians direct their efforts toward two different kinds of reality, one given by nature and the other instantiated in human practice. Rather, beginning with Galileo, modern science testifies *against* the very same kind of reality that history, beginning with Herodotus, testifies *for*: namely, the reality of appearances. Against the logic of the two-sciences settlement, Arendt's genealogy suggests that there has never been any such thing as "natural" science, properly speaking, for modern science begins with and is defined by modes of perceiving and speaking that disavow the reality of nature as it gives itself in appearances on earth.

The origins of modern science relayed by Arendt thus mark a tragic turn in both the history of truth-telling and the history of politics. What she sees as radically new—and radically dangerous—in modern science is a practice of spectatorship that *refuses* to receive earth's gifts of appearances and a preference for phenomena that withhold themselves from the naked eye. Unreceptivity to earth's gifts of appearances is politically pernicious because, as Arendt has helped us to understand, only those things that offer themselves to the senses in a surfeit of semblances have the capacity to separate and connect spectators, in the plural, from their various standpoints. Phenomena that are so shy as to require the interposition of an instrument to be called out of hiding cannot serve as an in-between; unlike earth's appearances, they cannot grow into a world. This is why earth-loss heralds world-loss, and why Galileo's glances constitute a political event of the first order. The image of Galileo with a telescope pressed to his eye serves Arendt as an icon of willful blindness to the diversity of sights, sounds, smells, tastes, and textures on earth's surface, and willful deafness to earth's law-like call to add political plurality to the natural diversity given on this planet.

Maker's Knowledge and the Fate of the Given

Critical though Arendt is of Galileo, her assessment seems to leave room for hope that subsequent innovators could build upon his triumphs in order to combat the despair of earth alienation, developing new sciences suited to augmenting common sense and reestablishing intimacy with earth's givens. But this is not how Arendt sees Galileo's descendants: on the contrary, she reads the history of science from the seventeenth to the twentieth centuries as a succession of all-too-successful movements to install "the mistrust of the given" as a cornerstone of modernity (*HC* 298). Above all, she perceives hostility toward earth's givens in the rise of maker's knowledge, the view that "man can at least know what he makes himself" (*HC* 282). Despite obvious and important differences in their methods and objects of study, the architects of maker's knowledge, as Arendt characterizes them, aligned in their desire to replace earth's givens with alternative realities of their own creation. On her analysis, the influence of maker's knowledge made hostility toward the given into a defining principle and governing ideal of the natural sciences and "the most general and most generally accepted attitude of the modern age."

By now it should be clearer why Arendt considers unreceptivity toward earth's givens to pose a threat to politics, the world, and common sense—and why her concerns should be taken seriously by proponents of world-building politics today. But why does she consider the embrace of *making* by scientists to be hostile toward givenness? Arendt herself has led us to recognize the important role of making in providing nature's ephemeral phenomena with a durable worldly context in which to appear, at once enhancing the realness of the given and amplifying its call to action, "the law of the earth." In coming chapters, I will work with and against Arendt's critique of science to uncover resources *within* the maker's knowledge tradition for fostering intimacy with the given and sustaining world-building politics. But first, to better understand why it is worth reclaiming these resources from the maker's knowledge tradition, I set out in this section to construe more specifically what troubles Arendt about the artificial realities engineered within this tradition. We will see that Arendt's concerns with maker's knowledge are not objections to artificiality per se. Rather, she is concerned with those versions of artifice that threaten to erase the alterity, difference, and diversity that earth's appearances give in such abundance. She rightly fears that scientific projects premised on the elimination of otherness stand to imperil both natural diversity and political plurality, even if (as subsequent chapters will suggest) she wrongly overlooks the potential for maker's knowledge to engender openness to the alterity of nature and to enhance common sense.

Arendt points to the introspective method of René Descartes as an early and formative example of maker's knowledge and a telling illustration of the risks of approaches to the study of nature that diminish and devalue givenness. Descartes is, of course, legendary for turning inward in pursuit of new and more certain truths—a received view of Descartes as a thinker of thinking that Chapter 3 will place into question. Arendt herself largely accepts the established view of Descartes as an icon of introspection, though she adds an important amendment: she casts his turn inward as a quest to renew both truth and realness, and to restore the connection (severed by Galileo) between them. She points to Descartes as the first to recognize the apocalyptic consequences of Galilean earth alienation: his "reflection on the *meaning* of certain scientific discoveries . . . destroyed his common-sense trust in reality" (*LM* 52). In Arendt's modification of the familiar characterization, Descartes not only shared the long-standing suspicions of the philosophical tradition that the senses cannot disclose what is true; he also shared Galileo's novel suspicion that the senses cannot disclose what is real. Descartes responded to the crisis of Galilean earth alienation, Arendt argues, by rejecting earth's gifts of appearances altogether. Although Arendt sees Descartes' method of radical doubt as a way of rejecting all sensuous evidence of the outer world, she also sees that Descartes did not give up on reality altogether. Rather, he sought to construct a new and more secure version of reality within the corners of his mind, as an alternative to the sights, sounds, and smells that are given by nature and preserved by other forms of human artifice.

Arendt emphasizes that Descartes's doubt was both destructive and productive of reality: "If everything has become doubtful, then doubting at least is certain and real" (*HC* 279). The Cartesian thinking "I" feels its own indubitable realness with every thought, even (especially) the self-annihilating thought of its own nonexistence. Even as doubting works to void the assurances of common sense that appearing things are real, it also establishes the reality of the thinking "I." Arendt thus casts Cartesian doubt as both an imitation and a displacement of common sense: Unlike trees or tables, which derive their realness from the presence of multiple spectators, the *cogito* derives realness from *self*-evidence. As the introspective "I" reveals itself *to* itself, it both gives and receives evidence of its own existence without appearing to or involving anyone else. Or so Arendt holds: upon returning to Descartes in Chapter 3, we will see that Arendt, like many of Descartes's readers, has overlooked his commitment to crafting a public, worldly appearance for the otherwise private act of thinking. But Arendt's critical point about introspection remains important, even if the reading of Descartes she uses to develop this point is questionable. She is concerned that, unlike the world of appearances, there is only room for one within the ersatz world of the mind. Engrossed in private introspection, "man is confronted with nothing and nobody but himself" (*HC* 280). Arendt's understanding of the Cartesian *cogito* is both simplifying and clarifying, warning us against scientific projects that would shore up the reality of the mind by denying the reality of the common world.

From this Arendtian perspective, Descartes's hostility toward the given was heightened, not mitigated, by his aspiration to know natural phenomena outside the bounds of his mind as intimately as he knew himself. Distrusting his senses, Arendt argues, Descartes discovered a way to circumvent them through mathematics: "Descartes' analytical geometry . . . succeeded in reducing and translating all that man is not into patterns which are identical with human, mental structures" (*HC* 266). On Arendt's reading, the techniques of algebra enabled Descartes to convert "objective reality into subjective states of mind" and to manipulate these subjective shadows of appearing objects without, as it were, actually touching or otherwise perceiving the appearing phenomena from which they were derived (*HC* 282). According to the maker's knowledge logic that Arendt attributes to Descartes, these artificial mental approximations of nature's givens could thereby be known as self-evidently as the "I" knows itself. In other words, Descartes promised a method of knowing nature that would enable the knower to remain utterly unreceptive to the appearances through which nature presents itself.

Of course, mathematical techniques did enable Descartes and his followers to calculate nature's mechanical properties with great precision and, on the basis of this knowledge, to intervene to reshape the face of the earth—technoscientific developments that troubled Arendt, like many others in her generation. But her gravest worry about rationalism's influence on the natural sciences—and the most original insight of her critical appraisal of science in a Cartesian vein—is that this version of intimacy with nature serves to deepen the modern problem of estrangement from otherness, earth's most valuable gift to politics. As she puts it, "The 'seen tree' found in consciousness through introspection is no longer the tree given in sight and touch, an entity in itself with an unalterable identical shape of its own" (*HC* 282). The mathematical alchemy that dissolves an objective, perceptible tree into a subjective state of mind cannot absorb the tree's irreducible *alteritas*—the distinctive shape, color, and texture that sets it apart from all other trees and all other appearing phenomena and constitutes its contribution to earth's diversity. Such a method makes every tree into a mirror that reflects the "I" back to itself, preempting any confrontation with otherness. "It is impossible," Arendt writes, "to reach out from the mere consciousness of sensations . . . into reality with its shapes, forms, colors, and constellations" (*HC* 281). Descartes's so-called natural science is, on her view, just another version of "the playing of the mind with itself"; its truths are as self-evident as the *cogito, ergo sum* only because they are equally narcissistic (*HC* 284).

Knowledgeable though they may be, Arendt warns, scientists who depend on introspection instead of common sense to tether them to reality will remain unable to grasp the lesson that nature's strange varieties of shapes and colors are meant to teach us: namely, earth's law-like edict to protect and augment its diverse appearances through human work and action. Unlike makers of tables, books, or houses, who preserve and enhance the distinctiveness of the trees they harvest for wood or pulp, according to Arendt, the makers of mental artifice devalue and disappear the otherness of nature when they convert its phenomena into immaterial thought-things. From the political perspective adopted by Arendt, the problem with the artificial reality of the *cogito* is that, unlike the material, appearing things that it disappears and replaces, it cannot be common. The ersatz world into which Cartesian method fits us is, in words Arendt borrows from Whitehead, " 'the outcome of common-sense in retreat' " (*HC* 283). She elaborates: "Common sense, which once had been the one by which all other senses, with their intimately private sensations, were fitted into the common world . . . now became an inner faculty without any world relationship. . . . What men now have in common is not the world but the structure of their minds, and this they cannot have in common, strictly speaking" (*HC* 283). What is often misleadingly referred to as "common sense" after Descartes is located within man in the singular, not between men in the plural. No matter how similar our thoughts become, however, Arendt insists that we do not have our minds or their products "in common." What Arendt means by "common" in this context, as elsewhere, is a connection forged across difference: we have a world in common insofar as we each perceive different aspects of the same thing. Unlike the mind, the body-bound senses need not be "merely" private or inner; insofar as they mutually relate us to things that are outside of and other than each of us, they are public and common. "Common sense," as Arendt reclaims the term, exploits the alterity of appearing things in order to constitute connections among plural spectators. As she has shown us, the constitution of a common world depends on both the alterity of appearing things and the plurality of sentient beings. Both of these conditions of possibility for commonality are given on earth; neither is to be found in the private spaces of the introspective self.

Arendt's story of Descartes's refusal to receive these gifts thus further embellishes her point that the earth and the world mutually condition and enable each other's reality. Precisely because the earth gives itself as a bewildering abundance of semblances, it calls together heterogeneous spectators and enables them to have a world in common. And even as these spectators depend on the alterity of earthly nature to have a world in common, the earth depends on their plurality to secure its reality. Earth's givens are not self-evident: they evince themselves in a context, not in a vacuum, to those beings who are fit—and willing—to perceive them. Arendt's critique of Descartes thus opens into a wider warning about the political costs of natural sciences that make the mind into an artificial sanctuary from "the stimulation—or rather, the irritation—of the senses by objects other than itself" (*HC* 283). Rather than reconnecting disillusioned moderns to earthly nature, under Descartes's guidance, she writes, "the modern *reductio scientiae ad mathematicam* has overruled the testimony of nature as witnessed at close range by human senses" (*HC* 267). Versions of maker's knowledge that treat the testimony of nature as an unwanted irritation may teach us new truths, but they make it harder to hear earth's testimony in favor of plurality, its injunction to reveal ourselves in word and deed.

The Cartesian internalization of the activity of making is not the only version of maker's knowledge that Arendt believes to have pitted the project of knowing the earth against the project of building a common world. She even characterizes the history of the scientific experiment as a story of hostility toward the otherness of earth's appearances. Arendt does not deny that the experimentalist "depends on appearances, whether, in order to find out what lies beneath the surface, he cuts open the visible body to look at its interior or catches hidden objects by means of all sorts of sophisticated equipment" (*LM* 24). But on Arendt's analysis, the appearances on which experimental scientists rely are *not* given by nature. Rather, they are forced to appear by scientists "laying bare what appearances themselves never show without being interfered with." Arendt resituates experimental science as a practical application of the Cartesian maker's knowledge orientation directly to the earth's appearing phenomena and a pernicious retrenchment of modern hostility toward givenness.

Although experiments ultimately depend on appearances, in Arendt's unusual characterization they are designed first and foremost to disappear both the appearances given by nature and the presence of other, more receptive human spectators. As the anatomist's knife exemplifies, experimental equipment "deprives" material phenomena "of the exterior properties through which they show themselves to our natural senses" (*LM* 24). From Arendt's vantage point, laboratories are means to strip away the perceptible surfaces of scientists' objects of study in order to probe for the deeper, less overt causes or grounds of appearance. The very walls of the buildings that house experiments circumscribe a space of disappearance, "the unseen quiet of the laboratories," at once shielding insiders from sights and sounds outside and preventing outsiders from seeing what is exposed within (*HC* 324). Arendt casts the lab, like the Cartesian *cogito*, as an "escape" from "the primacy of appearance" in the world (*LM* 24). Within its walls, mechanical and mathematical instruments serve to occlude whatever would otherwise readily disclose itself to the unaided senses and to coax into view whatever is too shy to show itself. As Arendt portrays it, the mode of spectatorship distinctive of the experiment compromises the appearance of the visible, causing the invisible to appear in its stead.

The peculiar danger she sees in these experimental phenomena is not their artificiality per se, but their parsimony: whereas the earth gives evidence of its existence freely to all sentient creatures, experimental evidence is taken by force of human ingenuity and instruments. In this respect, although the products of scientific experimentation depend on appearing phenomena, they "are *not* phenomena, appearances, strictly speaking."[25] While an appearance, strictly speaking, is meant to be received by the sense organs of sentient creatures, experimental products often go unseen even by their creators: "We know of their presence only because they affect our measuring instruments in certain ways, and this effect, in the telling image of Eddington, may 'have as much resemblance' to what they are 'as a telephone number has to a subscriber.'"[26] As Davide Panagia correctly surmises, Arendt finds "the 'more real,' or hyper-real . . . registered by hyper-sensitive machines" to be troubling because this alternative reality "does not have an apparitional character that humans can engage."[27] Unlike such quintessentially artificial human works as tables or houses, the artifacts of scientific experiments are *more* difficult for human spectators to perceive than the natural givens they replace. The experiment, as Arendt describes it, turns the activity of making on its head: instead of stabilizing nature's ephemeral appearances, it dissolves them into streams of data that can only be perceived by the most hypersensitive of machines.

In Arendt's view, these streams of data cannot convey the otherness of the appearances they replace, even to the scientists who build and read the instruments used to generate and collect them. As in Cartesian introspection, "here again" in experimentation, "we deal only with the patterns of our own mind, the mind which designed the instruments and put nature under its conditions in the experiment . . . so that wherever we search for that which we are not, we encounter only the patterns of our own minds" (*HC* 286–87). Where Descartes withdrew from nature's appearing things into the quiet of his mind, the experimental scientist withdraws *with* nature's appearing things into the quiet of the lab in order to contort them into reflections of his own intentions. Arendt casts the laboratory apparatus as a means of extruding the mind's patterns, purposes, and even its invisibility into a sheltered corner of the world. The establishment of the laboratory's liminal reality of not-quite-appearances is premised, she suggests, on extinguishing alterity, "that which we are not," from the objects of natural scientific inquiry. Surrounded by a nonappearing reality of their own making, these scientists "obviously only deal with a hypothetical nature" (*HC* 287). Just so, Arendt considers their knowledge of their laboratory constructions to be as narcissistic and unworldly as the Cartesian *cogito, ergo sum*: "Instead of objective qualities, in other words, we find instruments, and instead of nature or the universe—in the words of Heisenberg—man encounters only himself" (*HC* 261). Alienation from the nature given outside of the laboratory's doors is the price of too much intimacy with experimental apparatuses, as it is the price of too much intimacy with oneself.

Conclusion

Arendt's critique of modern scientific earth alienation thus puts on the table a set of political dilemmas that often go unnoticed by thinkers of the politics of science and exceed her own warnings about the risks of technoscience to human life. She helps us to recognize the potential for the very scientific enterprises devoted to knowing nature to estrange us from its reality, and to understand why such alienation constitutes a crisis for the world and for life in common. My readings of Arendt's writings on science suggest the need for deeper inquiry into the ways in which the natural sciences shape our political relationships with one another by constraining the conditions under which we encounter the alterity of nature. Arendt presses political theorists to evaluate the sciences not merely in terms of their validity or practical utility—concerns that I follow Arendt in bracketing—but in terms of their implications for the political project of building and sustaining a common world.

Arendt's own appraisal of the significance of the natural sciences for world-building politics is dark indeed. She believes that the rise of the modern sciences and their organization around principles of maker's knowledge were detrimental to common sense, our capacity to receive and guarantee sensuous evidence of earth's reality and to fit into a common world. On her analysis, the various guises of the natural sciences through history are but various means of diminishing perspectives and destroying appearances: Galileo's telescope, for her, afforded mechanical sanctuary from plurality and alterity, much as the walls of today's laboratories delimit who may participate in doing violence to the visible within, and introspection continues to erect virtual walls against encounters with nature's givenness. In these spaces of disappearance, she thinks, reluctant realities may be drawn forward, measured, analyzed, and mastered with the aid of sophisticated instruments and mathematical techniques. But on her analysis, the artificial realities constructed by the sciences are alternatives to, not parts of, the worldly reality received and guaranteed by common sense. Arendt has no confidence that mathematical formulae and experimental data could serve as an in-between to separate and connect men and women, in the plural, because they lack the propensity to "*appear* and . . . to be seen, heard, touched, tasted, and smelled," a propensity that *all* things in the world, natural and artificial, "have in common" (*LM* 19). Nor, in her view, could these scientific constructs serve the special political role she ascribes to natural things: namely, to remind us of all that we are not, even as they induce each of us to reveal who we are. Arendt portrays the natural sciences in tragic terms, as a disruption in the mutually sustaining reciprocity between earth and world and thus as a peril to natural alterity and political plurality alike.

Arendt is not wrong to point to the history of the natural sciences as a site where modernity's twin syndromes of earth alienation and world alienation have recurrently emerged and from which they have spread, even if—as coming chapters will show—her portrayal elides some of the messier, more ambivalent and even promising imbrications between science and politics.

There is perhaps no more telling—or tragic—illustration of the legitimacy of Arendt's concerns about scientific earth alienation than the struggles of scientists who wish to reenter the space of appearances to share the results of their researches with wider audiences. Often enough, these well-meaning would-be truth-tellers find that words fail them when it comes to describing what they have seen in the lab or in introspection. Arendt helps us to understand this predicament in light of the connection she helps to reveal between common sense—the ability to receive nature's testimony in favor of its own existence—and the power of speech to testify to the reality of nature's appearances. She warns that scientists who refuse earth's gifts of appearances are "forced" by this refusal to also "renounce normal language, which even in its most sophisticated conceptual refinements remains inextricably bound to the world of the senses and to our common sense."[28] Indeed, to Arendt, "The reason why it may be wise to distrust the political judgment of scientists qua scientists is not primarily their lack of 'character' . . . but precisely the fact that they move in a world where speech has lost its power" (*HC* 4). From this perspective, scientific jargon, far from a correctable verbal tic, reflects a much deeper injury to human speech. The replacement of earth's givens by artificial realities that do not present themselves to the senses may compromise the power of human speech to present objects of observation for political deliberation. And, as Arendt puts it, "wherever the relevance of speech is at stake, matters become political by definition" (*HC* 3).

So alongside Arendt's better-known warnings about the dangers of technoscientifically enhanced powers to act into nature and radically alter its processes, her writings on science, read closely, also alert us to the dangerous potential of the natural sciences to *diminish* the power to act by making it increasingly impossible to speak for nature. This trade-off is what she has in mind when she repeatedly warns that, by virtue of science, "man can *do*, and successfully do, what he cannot comprehend and cannot express in everyday language."[29]

Indeed, the scientifically enhanced power to act into nature beyond the walls of the mind or the lab is, for Arendt, one key reason that scientists are not the only ones to find that they are unable to encounter and speak for nature's alterity. From her perspective, the fact that "the world of the experiment seems always capable of becoming a man-made reality" endangers the otherness of earth even beyond the walls of the lab. When science is used to reengineer all of reality in our own image, this "unfortunately puts man back once more—and now even more forcefully—into the prison of his own mind, into the limitations of patterns he himself created," so that "the moment he wants what all ages before him were capable of achieving, that is, to experience the reality of what he himself is not, he will find that nature and the universe 'escape him'" (*HC* 288). Arendt helps us to see that sciences and technologies that erode the barrier between the world of the experiment and the world of appearances threaten to deprive all human spectators of contact with givenness and to render the earth not only uninhabitable, but unspeakable.

But Arendt does not help us to chart a way beyond the impasses that her description of the politics of science helps us to recognize. Her way of challenging the nature-politics opposition casts natural science as an antagonist to both earth and world and creates the impression that scientific aspirations to know nature are simply incompatible with political aspirations of building a common world, as though to choose one is to sacrifice the other. Although she helps us to see the conceptual and political insufficiencies of organizing our thinking about science around a two-sciences settlement, she stops short of developing a vision of natural science as a risky, but also promising, world-building practice in its own right.

In the next chapter, with the help of one of the most influential figures in the maker's knowledge tradition, Giambattista Vico, I begin to complicate Arendt's portrayal and develop an alternative account of natural science as both an extension of and a departure from world-building politics. My Vichian reappraisal of maker's knowledge is nonetheless informed and inspired by the key Arendtian insight emphasized in this chapter: namely, that the fate of the common world and the extrasubjective web of relationships that holds it together depends not only on our willingness to reveal ourselves in action, but also on our willingness—and ability—to encounter "that which we are not" in nature. Arendt's name for that ability, common sense, must be respected and augmented by natural science if it is to play a productive role in building and sustaining the common world and in speaking for nature in political contexts. For Vico to convince us that a science organized around maker's knowledge has the potential to play that role, we will need to find an ethos of receptivity toward the given within his account of scientific practices of making and knowing.

CHAPTER 2

Vico's World of Nature

Giambattista Vico is primarily remembered (when he is remembered at all) for his antipathy towards René Descartes. Vico lamented the neglect of what he called "the world of nations" by scholars infatuated with Descartes's mathematical approach to physics, and offered his 1744 *New Science of the Common Nature of the Nations* as an alternative, historical approach to the study of politics.[1] Since that time, Vico's works have been saved from the brink of obscurity more than once by readers looking for a friend in the enemy of their enemy. For example, some prominent advocates of the contemporary two-sciences settlement, seeing behaviorism as a second coming of Cartesianism, enlist Vico as an ally to protect the autonomy of political science and history against encroachment by mathematical methods.[2] Hanna Arendt, for her part, disagrees strenuously with these friendly estimations of Vico: in her view, Vico's moves "to abandon the attempt to understand nature . . . and to turn instead exclusively to things that owed their existence to man" did not protect the world of politics against the most damaging aspects of the Cartesian legacy.[3] On the contrary, she sees Vico's embrace of maker's knowledge as a capitulation to Descartes's methods and a deepening of the hostility of modern science toward the earth as it gives itself in appearances. As we have seen, Arendt criticizes Descartes for repudiating common sense, the faculty through which we receive earth's appearances and fit ourselves into the common world. For Arendt, Vico's turn away from the science of nature only exacerbated the modern problem of estrangement from the alterity and diversity of earth's givens, at the expense of the common world and of politics.

These contending interpretations of Vico overlook what is most original and valuable about his rebellion against Descartes, because they both take for granted that Vico's turn to the man-made "world of nations" represents a repudiation of nature and of natural science. I will show in this chapter that

Vico did not accept the opposition between the given and the made that shapes the reception of his work by critics and admirers alike. Rather, I argue, Vico offers a novel vision of politics as the historical creation of a world of relationships *among* the diverse phenomena given by nature. Wind, water, plants, animals, and the natural processes that push human and nonhuman bodies through lifecycles and weather patterns are all, for Vico, historical phenomena, as much a part of the world of nations as any other political institution. When Vico tells us that it is possible to know the products of human world-building practices, he affirms the distinctive intelligibility of natural things that appear within the context of political relationships. My readings show that, far from rejecting nature in favor of politics, Vico stakes the authority of science to speak for nature upon scientists' ongoing involvement in the political practice of composing a common world from nature's disparate elements. I recover guidance from Vico's works to help us move beyond the oppositional ontology of the two-science settlement and toward a more productive alliance between natural science and world-building politics than Arendt herself thought possible.

Vico's novel twist on the maker's knowledge tradition avoids displacing the given with the made by redescribing making as an activity that is both creative and receptive. Like Arendt, Vico places common sense at the center of his account of world-building politics, and in key respects his conception of common sense anticipates her understanding of nature's role in inspiring political action and fostering a "web of relationships" among plural human spectators by appearing between them. But Vico's account of common sense evokes a deeper and more durable intervention by these spectators into the configuration of bodies, human and otherwise, in political space. On Vico's account, participants in common sense establish relationships among themselves *in and through* the act of constituting relationships among the disparate objects of their shared scrutiny. As Vico's imaginative genealogy portrays them, the founders and refounders of the world of nations contribute relation, measure, and meaning to things that are given, but not related, by nature, setting all manner of material phenomena into man-made, worldly constellations through their speech and action. Vichian common sense embodies both receptivity and generosity: through common sense, we take in sensuous evidence of the otherness of the given even as we give *to* the given a relationship to that which it is not. Much as a metaphor likens across difference, common sense, for Vico, sutures together diverse human and nonhuman beings into a common, hybrid world, at once natural and political, material and historical.

When Vico tells us that we can know the world of nations because we made it, he is neither excluding nature from the purview of science nor sanctioning what he, like Arendt, sees as Descartes's troubling hostility toward appearances and givenness. Rather, by affirming the products of common sense as the appropriate objects of scientific inquiry, Vico affirms the receptivity *and* the creativity of both politics *and* science. The only reality we can really know, Vico suggests, is the world of relationships made by humans through their long history of contingent encounters with natural givens.

As this gloss of the argument already suggests, Vico doesn't just disagree with Descartes's answers to methodological and epistemological questions about *how* we know what we know; more importantly, for our purposes, Vico insists on the need to raise a different, prior set of questions about *what* we may know and how it came to be a part of our world. Which phenomena are and are not intelligible or even perceptible and thus available to be observed, contemplated, or analyzed according to the dictates of this or that scientific method? Vico's genealogical excavation of the acts of making that precede and surround scientists' acts of knowing and his claims that these acts depend in intricate ways upon one another do not reduce epistemology to politics or science to common sense. Instead, he insists on the differences between knowing and making in order to better understand their imbrications, dependencies, alliances, and tensions. Vico helps me to bring a different emphasis to the study of natural science, one that focalizes moments in the political past that gave to nature's givens a place within the common world and rendered them susceptible to scientific inquiry, among other human purposes. By theorizing the rise of the sciences in his early modern period as *both* a continuation of *and* a break from history-as-usual, Vico carves out a unique critical vantage point on the political conditions and implications of scientific authority. On the one hand, he anticipates Arendt's concerns with figures like Descartes who would premise their authority to speak for nature on isolation from the broader context of political activity. But his genealogy casts such isolation as a corruption of the deep historical connections between making and knowing, not as the founding, defining feature of science. Vico's own new science models the possibility of founding the authority of science to speak for nature upon ongoing involvement in the recursive making and remaking of the world in which nature becomes legible.

My recovery of this Vichian approach to the politics of science begins by turning to the text in which Vico first articulates his understanding of maker's knowledge, *On the Most Ancient Wisdom of the Italians* (hereafter cited parenthetically as *AW*). I show how he reworks threads from the rhetorical tradition to suggest that practices of forging connections among disparate things foster the worldly conditions of possibility for experimental natural science. The second section then turns to Vico's story of the generation of the "world of nations" in the *New Science*. Contra received readings, I argue, this story highlights the capacity of human actors exercising common sense to remake nature and to know it as *part* of the humanized, political world. I show how history serves Vico both as a connecting link between common sense and natural science and as a means to critique modern science for betraying its historical precedents. Finally, I conclude by commenting on what it would entail today to emulate and repurpose the new kind of new science that Vico models in the *New Science* (hereafter cited parenthetically as *NS*).

Maker's Knowledge Redefined

Vico is typically remembered not for challenging the opposition between natural givens and political artifice, but for originating the argument that nature and politics are too different to be studied using the same methods. Isaiah Berlin, for example, reads Vico as an early advocate of interpretation in sciences "that pertained to men as such, as contrasted with that which they shared with the non-human world."[4] Similarly, H. P. Rickman sees Vico as a precursor to contemporary arguments in favor of the autonomy of the *Geisteswissenschaften* or human sciences: Vico showed "the nature of the human world" to be a "socio-historical reality, which is meaningful because it consists of expressions of men's thoughts, intentions, feelings, and valuations."[5] Vico is indeed an early advocate and adherent of interpretive, historical approaches to the study of politics. But he does not treat the reality studied by natural scientists as a world apart from politics, history, or meaning. On the contrary, Vico suggests that natural phenomena become available for scientific inquiry and observation insofar as they are incorporated into the intelligible, humanized world of nations. Vico complicates familiar distinctions between the meaningless natural world and the meaningful sociohistorical world by according ontological status to the relationships created from natural matter through human praxis.

Vico's argument begins by accepting the basic precept of the maker's knowledge tradition—that we can only really know what we make—while also complicating this precept's traditional epistemological cast.[6] In *On the*

Most Ancient Wisdom of the Italians, Vico stipulates, following the suggestion of ancient Latin etymology, that "*verum* (the true) and *factum* (what is made) are interchangeable" (45), so that "the true is precisely what is made" (46). Vico's so-called *verum-factum* principle asserts that the possibility of knowing an object rests upon the conditions of its generation: only those entities that were made are intelligible, and only to their author. As Max H. Fisch emphasizes, *verum* or "true" in this context "means the true, not the truth, and its plural *vera* means not the truths but the trues or intelligibles."[7] To make a thing is not yet to know it, but to render it intelligible and thus to make it available for a wide variety of human purposes: to be perceived, grasped, used, cared for, spoken of, measured, evaluated, and so on. Vico recognizes that knowing is just one way of partaking of the familiarity engendered by the act of making. He recasts science as part of a continuum of human activities that reap the advantages of the special legibility of the made to the maker. To equate the true with the made is to begin to break the epistemological equation between truth and validity and to draw science closer to the practical wisdom or tacit knowledge cultivated through day-to-day interactions with material objects of necessity or pleasure.

In the *New Science Concerning the Common Nature of the Nations*, with his *verum-factum* principle in mind, Vico argues that political reality is uniquely susceptible to scientific treatment. If we reflect upon the fact "that the world of civil society has certainly been made by men," we can only "marvel that the philosophers should have bent all their energies to the study of the world of nature, which, since God made it, He alone knows; and that they should have neglected the study of the world of nations, or civil world, which, since men had made it, men could come to know" (331).[8] Vico likely has in mind passages in the *Discourse on Method* where Descartes vows never to apply his methods to the study of historical institutions, lest his science be viewed as a reformist political project. From Vico's perspective, this omission is astonishing: in the name of political neutrality, Descartes and his followers would cut themselves off from the very man-made reality that is most accessible to human observation and isolate the act of knowing from the acts of world-building that engender intimacy with this reality. The man-made, civil world is uniquely accessible to scientific scrutiny, according to Vico's way of thinking, precisely because it is historical, created through human political activity. Unlike this intelligible "world of nations," Vico argues, the "world of nature" is unintelligible, and thus unknowable, because it was made by God, not by men.

The passage makes clear that readers like Berlin and Rickman misconstrue Vico when they enlist him to support the two-sciences settlement. Far from espousing the complementarity of two sciences, suited for two distinct worlds, Vico thinks there is only one reality available for scientific investigation: the man-made, civil world. According to Vico's version of maker's knowledge doctrine, matter and motion as they were created by God are illegible to mere men—not suitable objects for science at all. Yet Vico does not simply reverse Descartes to reject the natural sciences in favor of a historical science of politics. Rather, he undertakes a novel inquiry into natural science's conditions of possibility: if the "world of nature" is unintelligible and unavailable for human purposes such as science, is there a nature that we *can* access and even come to know because it is built into the meaningful world that men have made? Through which actions and powers might human beings contribute to the generation of material reality? Taking up these questions, Vico ultimately suggests that natural science and political science are *both* possible, and for the same reason: because human praxis reworks the given elements of the "world of nature" into constitutive nodes within the world of nations.

Taking up the question of how mere humans could make and know physical reality, Vico considers and rejects Descartes's turn to mathematics, which Vico (much like Arendt) reads as a turn away from material givens in favor of immaterial mental constructions. On the one hand, Vico affirms the special intelligibility of the "world of shapes and numbers" to the mathematician, who "defines the names themselves, and on the model of God . . . creates from no substrate the point, the line, and the plane, as if from nothing and as if they were things" (*AW* 50, 51). The ex nihilo creation of this geometric world renders its immaterial figures as immediately transparent and familiar to its human makers as material phenomena are to God: "Just as divine truth is what God sets in order and creates in the act of knowing it, so human truth is what man puts together and makes in the act of knowing it" (*AW* 46). As Vico portrays it, the divine "act of knowing" emulated by the mathematician is an act of giving the rule to reality. A geometrical demonstration, on this view, is no mere proof—the mathematician's words at once shape reality, take its measure, and represent it, all in one breath. To know by making, on this view, is closer to divine omnipotence than it is to epistemological accountability to standards of validity.

But if the *verum-factum* principle helps to explain the distinctive, almost incontestable authority of a geometrical demonstration, Vico concludes, it also underscores the difficulty of establishing anything like omnipotence when it comes to nature. "Of course, the physicist cannot truly define things, that is, he cannot assign to each its own nature and thus truly make it, for that is God's right but is unlawful for man" (*AW* 51). Insofar as physical reality is authored by the divine word alone, the *verum-factum* principle teaches that it is only intelligible to the divine understanding. Vico compares the special familiarity that God and geometers have with the worlds they respectively author to the legibility of a text to a literate reader: "Just as *legere* (to read) is applied to one who combines the written elements of which words are composed, so *intelligere* (to understand) may be the combining of all parts of an object from which its most perfect idea may be expressed" (*AW* 46). The so-called book of nature is, in this sense, composed in a script that we cannot read or write. For there to be any such thing as a science of nature, there must be some way that our practices leave a signature in this book. Vico's animating question—in what respects does the human creature participate in creating nature?—is heightened, not resolved, by contemplating mathematicians' successes in making and knowing geometrical abstractions.

It might seem as though Vico means to resolve this problem by affirming the successes of experimental methods in the natural sciences. Vico was both a critic of the Cartesian reduction of science to mathematics and an enthusiastic observer of the experimental culture of early modern Naples and other European cities.[9] In *On the Most Ancient Wisdom*, he endorses observation and induction as the best methods to study nature's causes (103) and casts his maker's knowledge doctrine as "a handmaid for experimental physics . . . for in this metaphysics we regard as true in nature that to which we make something similar through experiments" (109). Experiments create imitations of effects observed in nature. According to the *verum-factum* principle, these experimental facsimiles of nature are "trues," more intelligible to the scientists involved in making them than to uninvolved bystanders, and more intelligible than nature as it is given prior to the experiment. Vico's embrace of experimental techniques as the best way to approach the physical sciences complicates prevalent views of Vico as a partisan of the human sciences and affiliates him with the lineage of pragmatism, as Fisch and others have argued.[10]

But if experiment were the end of Vico's story of human involvement in the making of nature, as these pragmatist readings largely assume, it would remain difficult to understand his confidence in the intelligibility of nature to the experimentalist. Indeed, Arthur Child considers this confidence unjustified: according to Vico's own presuppositions, "experimental physics is not, evidently, knowledge at all," because we do not, in experimenting upon natural things, create them.[11] Rather, as Arendt was keen to remind us in the previous chapter, experiments create artificial results by destroying or defacing nature as it is given. Even Fisch thinks that experiments yield only "quasi-truth," given that, for Vico, nature prior to the experiment is "none of our making."[12] If Vico does believe that natural phenomena are purely given by God and are "none of our making," then scientists could have nothing to say about these phenomena on the basis of their experimental results. Analogies between experimental and natural phenomena would be groundless, according to the logic of his *verum-factum* principle, and efforts to "read" the outcome of an experiment as evidence of nature's causes would be incoherent. How can we know that our knowledge of a man-made experimental phenomenon is a reliable guide to an original made by God alone and intelligible only to him? Both the authority and the validity of scientific experiments seem to presuppose what needs to be established: namely, that nature is intelligible in the first place because we were involved in making it. The successes of the experimentalists admired by Vico indicate, but do not themselves account for, a prior contribution to nature by man, raising once again Vico's larger question: Where do we see human beings acting to set in order, put together, and create the natural reality that experimental scientists take apart, analyze, and recreate?

Unlike some of his readers, Vico's investigations of *verum* and *factum* in nature do not stop with the creativity of the experiment. Rather, he makes the tantalizing and—as he is aware—perplexing suggestion that the rhetorical tradition offers promising avenues for understanding the intelligibility of nature to human spectators. As he puts it, "From the same founts from which well-equipped speakers spring, there also might come forth the best [scientific] observers" (*AW* 101). It seems that Vico thinks that excellent speakers will also excel at reading the reality of nature. In particular, he argues that "physics can be advanced . . . through the cultivation of *ingenium*," the traditional faculty for speaking in metaphors—anticipating that this position will be "surprising to anyone who is concerned with method" (104).[13] Vico suggests that partisans of both experimental and geometrical methods have overlooked the role of rhetorical aptitudes in the physical sciences. In what respect is metaphor, typically seen as a means of persuading—if not of distorting or misrepresenting—actually a crucial part of the story of how physical reality is made by and may be known by human observers?

Vico begins to recast *ingenium* as an enabling condition for scientific observation by turning our attention from the metaphor's afterlife as a persuasive saying to the moment of its birth, when a hitherto unrecognized similarity is first articulated in speech. He defines *ingenium* as "the creative power through which man is capable of recognizing likenesses and making them himself" (*AW* 102). *Ingenium* is a power of invention, enabling speakers of human language to make connections between similar things. According to the *verum-factum* principle, such connections are intelligible to those who participate in their genesis. Notice that Vico acknowledges the *ingenium* of both speakers and listeners: to recognize a likeness evoked by a speaker demands a bold leap from particular to particular to discover what they have in common: "The sharp men are the ones who are able to find a likeness or ratio between things very different and far removed from one another, some way in which they are cognate, or who leap over the obvious and recall from distant places the connections appropriate for the things under discussion." The rhetorical alacrity that allows "sharp" speakers and listeners to connect disparate things is also, Vico claims, the "fount" of scientific observation in the physical sciences. Putting his own *ingenium* on display, Vico catches sight of a likeness between the otherwise disparate scientific and rhetorical tasks of discerning hard-to-see resemblances. Rhetoric, in his view, is both the art of persuasion and the fount of keen observation.

Concomitantly, to observe is not merely to receive stimulation from the thing observed, in Vico's view. Indeed, to receive itself requires the active, creative participation of the observer's sensorium: "We make the color of things by seeing, flavor by tasting, sound by hearing, and heat and cold by touching" (*AW* 93). Sensation is a creative and a receptive mode of experience. Those who observe objects, like those who listen to speech, must actively participate in making connections between disparate particulars to grasp the reality presented to them. On this view, the difference between a "mere" sensation and a sharp observation is a matter of the *ingenium* involved. Speakers who are especially skilled or practiced at articulating acute rhetorical connections make for especially acute scientific observers because they are able to discern resemblances that others might overlook. Good scientific observers are involved in reworking the world, contributing relationships between unrelated things and recognizing the relationships contributed to the world by others. As Vico scholar Ernesto Grassi puts it, "insofar as metaphor has its roots in the analogy between different things and makes this analogy spring into sight, it makes a fundamental contribution to the structure of our world."[14] Based on the *verum-factum* principle, this human contribution of color, flavor, likeness, and ratio to things makes them newly intelligible to human observers and hospitable to scientific experiments.

Vico's recuperation of *ingenium* thus provides a backstory to explain how the objects of natural science came to be *vera* or "trues," susceptible to human knowledge, in the first place. According to this story, the stuff of nature becomes available for human observation and experimentation as human language users weave things that God made ex nihilo into a meaningful web of rhetorical relationships. Without these man-made relationships to hold it together, nature would be an illegible cacophony of radical diversity, particularity, and flux: "In nature," Vico writes, "there is neither straightness nor unity, nor sameness nor rest" (*AW* 92). Sameness is a rhetorical construct, given by "the sharp men" to the otherwise disparate entities of the God-given world of nature. Even the self-sameness of a single thing is not part of the natural endowment given to the world by God. Echoing Heraclitus, Vico writes: "This is the life of things, just like the river that seems the same, yet is forever the flowing of different waters" (*AW* 82).[15] Things as they are given by God are not self-same objects, but flows of infinite difference. Yet humans can observe the "same" river through time and even pose and answer causal questions about its fluctuations. From a Vichian perspective, the creation of sameness from difference, the work of metaphor, is a prior, enabling condition for the investigation of causes from effects, the work of induction or metonymy.

In Vico's version of maker's knowledge, rhetoric rivals geometry as a way for human beings to approximate the divine power to make and know. But where geometers attain their capacity to make and know lines and planes at the expense of leaving material reality behind, on Vico's account, those who deploy *ingenium* rework matter into artificial, but still material, metaphorical networks. Vico portrays God and man as coauthors of a world that is both physical and rhetorical. According to the *verum-factum* principle, this hybrid world, not the "world of nature" made by God, is the proper object of the physical sciences. Of course, Vico considers *ingenium* a paltry power compared with the ex nihilo power to make everything from nothing that he attributes to God: "We cannot understand the limitless and formless, though we can think about them. As we say in Italian, 'One can keep on picking things up, but never get them all together'" (*AW* 77). But what human beings lack in creative potency we may make up for with persistence and receptivity. Over the vast epochs stretching from the most ancient Italians cited in Vico's title to his contemporary colleagues in the experimental societies of Naples, human beings gradually made a world by picking up and putting together starting materials received from a limitless power, which put them here for us to find. For Vico, this power is God. This long history of rhetorical praxis stands mostly forgotten in the background of experimental science, practitioners of which remake, in their experiments, things that were previously put together by other, prescientific, rhetorical means. The world known by physicists is a product of history, on which the sciences quite literally build.

In Vico's hands, *ingenium* is no "merely" rhetorical faculty for adding ornamental flourishes to expository language. Rather, he resituates rhetoric at the center of the world-building practices through which human actors contribute sameness to the pure difference given in nature. Vico's suggestion that human rhetorical praxis contributes relationships to the reality given by God and studied by natural scientists can help us to entertain a more radical departure from the two-sciences settlement than readers like Fisch allow. Fisch's emphasis on Vico's account of the creativity of the experiment challenges received views of Vico as a thinker of the human sciences with little to say about the natural sciences. But Fisch still compartmentalizes Vico's humanist commitments—his emphasis on the creativity of rhetorical praxis, which reaches fruition in the *New Science*—from Vico's scientific commitments. Such readings leave us with the impression that the event of the experiment is the only encounter with nature that matters in the constitution of scientific authority. But Vico's turn to *ingenium* already gives us reason to question whether the authority of natural science is reducible to the validity of experimental procedure. His rhetorical twist on the maker's knowledge tradition suggests that both the felicity of experimental practice and the authority of scientific speech depend on the practitioner's relationship to wider scenes of rhetorical praxis and sensuous experience, past and present. Scientists who speak for nature benefit from and add to a long history of world-building practices, over the course of which countless human beings have woven a world of relationships from the infinite alterity of nature created by and received from God.

Vico tells his version of the history of this man-made, material world in *The New Science of the Common Nature of the Nations*. As the next section shows, his story redefines both politics and science by casting the enactment of common sense as the point of origin for both political reality and the authority of the natural sciences. But to grasp this aspect of Vico's intervention into the science-politics relationship, we will have to be unusually attentive to the place of nature in Vico's account of political founding.

Nature and the Genesis of Politics and Science

Vico's *New Science* is a beguiling text; its baroque complexities discourage many would-be readers while enticing some with the opportunity to find just about any lesson lurking in its intricacies. The only thing that most readers of the *New Science* seem to agree about is that Vico repudiates natural science in its pages.[16] For instance, Antonio Pérez-Ramos sees Vico's *New Science* as a "new era" for the maker's knowledge tradition: where previous "debate was predominantly centred on the problem of knowledge as manifested in the confrontation of man and Nature," in this text "the Baconian project and the ethos of the new scientific movement are left behind."[17] But such readings overlook the important role Vico attributes to nature in the genesis of the world of nations. In Vico's narration, the history of politics is largely the story of human beings' collective confrontations with nonhuman nature. Indeed, Vico suggests that it is only as *political* beings that human beings create, through history, both their own human nature and an intelligible nonhuman material reality. He recasts common sense as the power that draws together human and nonhuman bodies into lived political realities and enables human speakers to give voice to phenomena that cannot speak for themselves. As we will see, common sense conceived in these Vichian terms plays an integral role in the establishment of both political and scientific orders, and helps to explain the complex historical dependencies and antagonisms between politics and science.

Vico opens the question of the relationship between human political activity and nonhuman nature by imagining how confrontations with material reality would differ in the absence of politics, in a time before the world of nations was founded. His origin story begins by extrapolating from the biblical story of the flood: Vico conjectures that Noah's wayward sons and their descendants must have wandered alone through lush antediluvian forests, becoming "stupid, insensate, and horrible beasts" as they shed language, culture, faith, and reason (*NS* 374). How would isolated, dehumanized men and women such as these experience their natural surroundings? Vico imagines that these fictive, barely human creatures would lack, above all else, the capacity to relate disparate particulars—the faculty he taught us to call *ingenium* in *On the Most Ancient Wisdom of the Italians*. Their minds, unlike ours, "took things one at a time, being in this respect little better than the minds of beasts, for which each new sensation cancels the last one" (*NS* 703). According to Vico, nonhuman animals *cannot* make the color of things by seeing, flavor by tasting, and so on. Even the five human senses that we take for granted as our "natural" endowment, he suggests, actually emerge only in the context of our involvement in political community. The prepolitical human animal, unable to give sameness to nature, is shipwrecked upon the immediacy of each instant and encounters nature as an insensible torrent of unrelated particulars. Anticipating Arendt's account of earth alienation, Vico posits that the unraveling of human community erodes our ability to perceive the very reality of appearing things. Prior to the founding of the civil world, he thinks, people lived without any world at all, for human beings sacrifice both their humanity and their grip on reality when they sever ties with one another.

Conversely, as Vico tells it, the founders of the nations restored communal bonds among themselves—and thereby restored their fragile status as humans—by resuming their participation in the rhetorical practice of creating relationships among disparate things. According to his story, Noah's descendants were rescued from solitude by the intervention of a singular weather event: a tremendous thunderstorm, the first after centuries of floodwater evaporation. The sight of this "great effect whose cause they did not know" arrested the attention of several far-flung spectators and inspired their first collective act: "Because in such a case the nature of the human mind leads it to attribute its own nature to the effect, and because in that state their nature was that of men all robust bodily strength, who expressed their very violent passions by shouting and grumbling, they pictured the sky to themselves as a great animated body, which in that aspect they called Jove, the first god of the so-called greater gentes, who meant to tell them something by the hiss of his bolts and the clap of his thunder" (*NS* 377). The civil world began with the creation of an inaugural metaphor, a relationship of likeness between the disparate particulars of man's voice and body, on the one hand, and the sky's sounds and motions, on the other. By humanizing the sky, Vico's founders humanized themselves: obeying the meaning they gave to the sounds of thunder, they stopped wandering, settled together, and began to give meaning to one another's shouts and grumbles. A primordial political gathering, bound together by an early version of language, was reborn, in this story, through collective, embodied practices that bound men to the sky. Vico's fantastical origin story carries the message that natural phenomena are inextricable elements of the civil world's webs of meanings. Only within this civil world do weather events and other natural phenomena have something to say to human audiences, and only within this world do the shouts, grumbles, and eventually words of humans become legible as representations of nature. Natural things become both newly sensible and newly speakable at the birth of the world of nations.

Vico's origin story helps to contest the tradition of opposing natural matter to political meaning by foregrounding the ontological contributions made by the authors of politics to the reality of nature. As he describes it, from its earliest beginnings, the "world of nations" is composed of enacted metaphors that liken and gather together disparate human and nonhuman beings. For Vico, the world is built by giving relationships to things that are given, but not related, by nature, thereby receiving them into the civil world. As he says of the founders, "To all the universe that came within their scope, and to all its parts, they gave the being of animate substance. This is the civil history of the expression 'All things are full of Jove'" (*NS* 379). The figure of Jove does not just represent things, it fills them, at once adding to their being and incorporating them into a common, heterogeneous world. This sky-god is the first of many figures through which the founders of the nations received natural elements into their nascent collectives, according to Vico's detailed theogony in the *New Science*: they "gave living and sensible and for the most part human forms to the elements and to the countless special natures arising from them, and thus created many and various divinities" (*NS* 690). By divinizing, humanizing, and animating the elements, the makers of the nations made sense of nature, rendering it familiar, even conversant. By contrast with the insensible, unspeakable particulars of the "world of nature," the "world of nations" is a world of relations—its dense network of man-made, rhetorical bonds is legible precisely because they crisscross and defy the human/nonhuman distinction.

At these earliest moments in the history of the nations, Vico emphasizes, the first bonds among human and nonhuman beings were created not by escaping from the grit and gravity of physical embodiment into a weightless realm of linguistic abstractions, but by physically moving disparate human and nonhuman bodies into closer proximity. The very language through which the founders contributed animate substance to various parts of the universe was, according to Vico, not a system of linguistic signs but a "mute language of acts and objects" (*NS* 34). The civil world was generated through a rhetoric of deeds, as men heaved their own and other bodies into new relationships, incorporating them into the nascent human collective. Jove and other primordial metaphorical bonds between men and things were at once meaningful and substantial, imagined and real—what Vico calls a "real poem" (*NS* 217). Meaning is, at first, a material construct, created from natural elements through the alchemy of politics and legible to its human authors as the book of nature authored in a divine language could never be.

Vico's name for the power of concerted physical activity to create a world of relationships among men and things is "common sense," a power of "judgment without reflection, shared by an entire class, an entire people, an entire nation, or the entire human race" (*NS* 143). Common sense, as defined here and exemplified in the actions of the founders of the world of nations, is a pre-reflective, embodied, yet coordinated response to phenomena that present themselves to the senses of more than one spectator at the same time. Like the other five senses, Vichian common sense is both receptive and creative, at once opening the body to the alterity of that which is given and *making* sense of it. But where the eyes make color and the tongue makes flavor, Vico suggests, common sense makes commonality: the concerted efforts of plural spectators to receive the otherness of nature binds them to one another in and through the connections they mutually establish with their objects of common scrutiny. Like Arendt, Vico treats common sense as a faculty that is corporeal and shared, exercised in collective actions and passions, not private mental reflections. But where Arendt distinguishes the receptivity of common sense from the creativity of *homo faber*, Vico helps us to see that receptivity towards nature as it is given demands the creativity to constitute a relationship with otherness. In his origin story, the relationships created by the nations' founders spanned, without destroying or occluding, the distance and the differences between man and sky and among men, in the plural. Vichian common sense is responsible for creating a whole world of such relationships. Within this common world, the givens of nature become newly perceptible, intelligible, and ultimately even knowable. When Vico says that men cannot know the world of nature, he rules out the possibility of knowing nature as God does while suggesting the possibility that the natural sciences could pursue a different kind of knowledge by building upon the common sense that built the world of nations.

To some extent, Vico's account of the generation of this hybrid political world anticipates the contemporary new materialist literature, which similarly challenges traditional ontological oppositions between nature and politics. But new materialist thinkers like Jane Bennett and Bruno Latour, who seek to recover the agency, spontaneity, and loquaciousness of nonhuman things, would likely contest Vico's suggestion that material phenomena lack the agency to make themselves seen or heard and move themselves into meaningful configurations.[18] Although Vico does enlist a peculiarly lively and loud thunderstorm to cut through the paradox of founding and make the founders into makers, he does not let this aspect of his own story impinge upon his position that people, not things, are the founders of nations and the protagonists of history. Vico seems to assume that only a "world of nature" devoid of relation leaves room for the human creativity of common sense. The new materialist literature, by exploring the role of nonhumans in co-constituting politics, can help Vico's readers to challenge this assumption and recuperate more fully than he does the recursive loops between human creativity and the fecundity of nonhuman nature.

But Vico also foregrounds a question that this literature tends to neglect: how do human beings become capable of apprehending and making sense of the things in their midst? From a Vichian perspective, developing an improved account of the agency of matter, however worthwhile, cannot explain how things become perceptible and intelligible to human observers. For him, receptivity to the vibrations of things requires the creativity to establish a relationship with them. By insisting, anthropocentrically, that we only know what *we* have made, Vico encourages agnosticism about whether or not things really do speak or act for themselves: even if they did, only a god would have the ears to hear them. He redirects our attention to the story of the human constitution of politics in order to suggest that the only world that discloses itself to human observers is the one they come together to create. In Vico's account, the attunement between man and world is, first and foremost, a political accomplishment, established through the deeds of men, in the plural.

This Vichian amendment to the new materialisms also presses readers of Arendt to discover something that is occluded in her account of world-building politics: namely, the story of how the political world and its constitution in human speech and action develops into and enables the emergence of scientists as spokespersons for nature. Vico offers one way of telling this story in the *New Science*: he recasts the new sciences of his time, like quests for knowledge in the ancient past, as particularly consequential episodes *within* the historical drama of common sense. The *New Science* highlights both the distance and the continuity between the deeds that created the world of nations in the very beginning of history and the deeds that established the sciences at history's end: "The first peoples, who were the children of the human race, founded first the world of the arts; then the philosophers, who came a long time afterward and so may be regarded as the old men of the nations, founded the world of the sciences, thereby making humanity complete" (*NS* 498). Vichian history connects the past with the present not to disparage either the "children" or the "old men" of humankind, but to recast late-coming efforts to know the world as further developments in the history of the making and remaking of the world. For Vico, the common sense that created a corporeal metaphor like Jove is both historically distinct from and the historical condition of possibility for the scientific enterprise of seeking knowledge of the causes of thunder and lightning. As he puts it, "The sciences . . . sprang from the nations and from no other source" (*NS* 51). Vico's writings suggest that both the intelligibility of nature and the authority of scientists to speak on its behalf are founded on the common sense that built the nations.

But if the sciences are enabled by their roots in common sense, efforts to cut this root and separate the scientific act of knowing from the political act of making threaten to create crises of illegibility and ignorance. The *New Science* models a critical vantage from which it is possible to recognize the promise of science to enhance common sense while remaining wary of the risks of sciences that premise their authority upon devaluing or disdaining common sense.

Even as he connects them, Vico does not neglect the differences and tensions between making and knowing or politics and science. Common sense, as he casts it, is not science; it is not the way the world is known, but the way the world that may be known is made. Vico's founders exemplify the creativity of ignorance: "Man becomes all things by *not* understanding them . . . for when man understands he extends his mind and takes in the things, but when he does not understand he makes the things out of himself" (*NS* 405). Those who knew nothing about thunder were uniquely suited to make everything of it. Vico casts the absence of knowledge as a prelude to acts of world creation. These world-building practices, he argues, render things intelligible, paving the way for rational understanding.

Later, the mind that "takes in the things" absorbs, too, the history that made these things what they are. These things are not only "full of Jove"; they are full of the common sense of earlier generations, who made Jove out of themselves: "As much as the poets had first sensed in the way of vulgar wisdom, the philosophers later understood in the way of esoteric wisdom; so that the former may be said to have been the sense and the latter the intellect of the human race" (*NS* 363). Both the intellectual faculties on which scientific thinkers depend and the sensuous reality that they observe are historical artifacts, according to Vico, created through millennia of political praxis. The web of relationships woven in the past runs through the things that modern scientists contemplate, manipulate, and articulate. Indeed, the inextricability of scientific objects from historically fabricated meanings is precisely why these things are legible to a scientific gaze and speakable in a scientific language. The *New Science*'s long, epochal political history dramatizes the dependence of those who speak for nature upon those who acted to make it. The *vera* or "trues" that are the objects of Vico's version of maker's knowledge are a historical inheritance.

Vico's history also teaches that the activities of the natural sciences at the end of history are inspired by and spring from the ongoing world-building practices that surround scientists in the present. He illustrates this by drawing an analogy between his early modern present and more remote periods from the past: he turns to the rise of ancient philosophy to illuminate the dilemmas of founding a modern science. In antiquity, Vico argues, the capacity for abstract thinking was born through political tumult, as plebeians developed new "vulgar genera" to express their objections to aristocratic privilege. By means of these political and linguistic tumults, "the minds of the peoples grew quicker and developed powers of abstraction, and the way was thus prepared for the coming of the philosophers, who formed intelligible genera" (*NS* 460).

In Athens, for example, ancient philosophy and democracy arose side by side, Vico points out: "Now, because laws certainly came first and philosophies later, it must have been from observing that the enactment of laws by Athenian citizens involved their coming to agreement in an idea of an equal utility common to all of them severally, that Socrates began to adumbrate intelligible genera or abstract universals by induction" (*NS* 1040). The ancient democratic polis was, Vico writes, "the public school from which . . . the philosophers emerged" (*NS* 327). Socrates was prepared to become a philosopher by his involvement in and observations of this Arendtian space of appearances, where the citizen lawmakers of Athens acted to establish newly expansive relationships of equality among themselves. The experience of political equality gave Socrates practice at discovering what diverse particulars have in common. His distinctive feats as a philosopher—induction and abstraction—both extended and surpassed the creativity and receptivity of common sense. Imitating Socrates, Vico adumbrates a universal principle of his own to make the point: "The order of ideas must follow the order of institutions" (*NS* 238). The same principle applies to Vico himself and to other would-be scientific innovators: the founding of a new science can only follow from and build upon the instantiation of new political realities.

But if the act of knowing emerges from the act of building the world to which the knower belongs, then a decline in world-building practices presages a decline in the prospects for knowledge: "As the popular states became corrupt, so also did the philosophies" (*NS* 1102). According to Vico's genealogy, as citizens in later stages of antiquity eschewed or were excluded from democratic self-rule, their common sense was dulled by "the custom of each man thinking only of his own private interests" (*NS* 1106). For all the sophistication of their languages and ideas, the late-coming citizens of the ancient nations fell into an atomized condition akin to the plight of Noah's wayward descendants, unable to see what they had in common with one another or with the world of things between them. As the web of relationships began to fray, so too did philosophy: "Then came Aristotle and Zeno. The former taught the syllogism, a method which deduces particulars from their universals rather than uniting particulars to obtain universals. The latter taught the sorites, which, like the method of modern philosophers, makes minds subtle but not sharp. Neither yielded anything more notable to the advantage of the human race" (*NS* 499). Unlike Socrates, who learned in "the public school" of the polis to unite particulars to form new universals, Aristotle and Zeno took universals apart, dividing what earlier ages of political and intellectual practice had gathered together. Vico's criticism of Aristotle and Zeno makes them foils not only of Socrates, but also of the nations' earliest founders: where the first men were sharp but not subtle, their distant philosophical ancestors were subtle but not sharp. In Vico's view, Aristotle and Zeno lacked the *ingenium* to add something new to the history of human ideas and institutions. Instead, they devoted their acts of knowing to taking apart the web of relationships they inherited from the past.

Vico's concern with the late-coming philosophers of antiquity—and with Descartes, whom he casts as their modern corollary—is that they embrace and reinforce the political decline to which they owe their rise. In the case of the ancients, Vico sees the methods of Aristotle and Zeno as symptomatic of a broader way of life in which both philosophers and nonphilosophers lost their taste for common sense and gave up on creating a common world in public scenes of speech and action. These subtle philosophers, on his analysis, embraced and exacerbated the "civil disease" that Vico calls "the barbarism of reflection," an unraveling of the fabric of the civil world that he blames for destroying the ancient nations. In such an age, people cease to gather together to discover and create commonalities, and thus, for all their mental refinements, they "live like wild beasts in a deep solitude of spirit and will, scarcely any two being able to agree since each follows his own pleasure or caprice" (*NS* 1108). Isolated in mind, body, and spirit, the "old men" at the end of the ancient epoch began to resemble their opposites, the barely human beings who suffered such terrifying, lonely lives after the flood and before the advent of nations. As a result, Vico argues, the ancient nations fell, and human beings returned to their prepolitical beginnings, sacrificing their humanity in the process. "Obstinate factions and desperate civil wars" tore apart the worldly bonds that previous generations had constructed and turned "their cities into forests and the forests into dens and lairs of men." In Vico's hands, antiquity becomes an object lesson in the apocalyptic consequences of isolating the act of knowing from the act of making and of substituting philosophy or science for the common sense from which it springs.

History repeats itself, according to Vico's cyclical genealogy: centuries after the ancient nations fell, the modern nations were founded and passed through periods of theological, aristocratic, and popular rule, retracing the same broad pattern followed by the ancient nations before them. Vico offers precious little explicit commentary on the makers and knowers of the modern epoch in the *New Science*. But his cyclical narrative cleverly implies a likeness between the disorders that destroyed Athens and other ancient nations and the problems of his own time. *On the Most Ancient Wisdom* draws the connection more explicitly: there, Vico casts Descartes as a modern-day Zeno, warning that Cartesian method makes the mind "more subtle than acute" (103 and passim). The parallel suggests that when Descartes and his followers set out to study "the world of nature" in isolation from "the world of nations," they repeat ancient mistakes, to the peril of both modern science and modern politics. Indeed, Vico sees Descartes's new science as both politically pernicious and scientifically bankrupt for the very same reason: namely, its valorization and exacerbation of the decline of world-building politics, the very thing that destroyed the world of nations once before and threatens to do so again. By isolating the quest for knowledge from involvement in politics, Vico thinks, the methodological prescriptions of Descartes threaten to accelerate the unraveling of the civil world and return us to our beginnings as insentient, solitary, barely human creatures.

Conclusion

Vico, like Arendt, thus points to the rise of Descartes's approach to science as the beginning of the end of the common world. Vico anticipates Arendt's concerns that Cartesian method imperils the human connection with nonhuman nature and deprives human communities of the intimate entwinements with material phenomena fostered by common sense. Vico's critique of Descartes also anticipates Arendt's defiance of the oppositional logic of the two-sciences settlement: he, too, insists on the inadequacy of Cartesian method even in the study of nature. "Imported into physics," he writes, "the method is not useful for making new discoveries so much as for setting in order the discoveries we have already made" (*AW* 103). Much as Aristotle and Zeno failed to add anything notably new to human history, in Vico's view, their more recent Cartesian counterparts lack the *ingenium* and common sense to make or know anything. Vico's story of the barbarism of reflection, like Arendt's account of earth alienation, indicts certain powerful tendencies within the history of modern science for embracing isolation from the sensuous, collective practices through which we receive evidence of nature and constitute a common world. Vico and Arendt together warn that sciences that estrange us from our material surroundings and thus from one another endanger the web of relationships that holds the world together.

But Vico has a different understanding than Arendt of what it means to be alienated from nature. For Vico, nature as it is given is alien and unintelligible unless and until it is integrated into a world of relationships by receptive, creative human spectators. Until we act to forge connections among the givens of nature and thereby incorporate them into the human world, they cannot have the status of perceptible, legible objects of observation and contemplation, he suggests. Vico's origin story offers an unusually affirmative perspective on anthropomorphism: he suggests that the humanization of nature can be an act of generosity rather than narcissism. Vico's founders opened themselves to the frightening alterity of the unprecedented and unknown in nature in and through their acts of giving some of their own corporeal qualities to their material surroundings. The products of their deeds, metaphor-like connections between man and sky, human and nonhuman, both traversed and preserved the differences between the things thereby related. As they incorporated the nonhuman into their nascent human world, they became newly intimate and conversant with it. And in so doing, they drew closer to each other, discovering something common in something strange and renewing their own diminished humanity. Against this backdrop, Vico sees the specter of something like Arendtian earth alienation in the rise of sciences that *refuse* to participate in humanizing nature. He shares Arendt's sense that Descartes's version of maker's knowledge is distinctively hostile toward givenness, but he understands this hostility as reflecting too little, not too much, creativity on Descartes's part. From a Vichian perspective, the problem with science in the Cartesian vein is that it eschews and derides participation in the political practice of reshaping nature. His critique of Cartesian science points toward the possibility of a different kind of natural science—one that would inherit and rework relationships forged in the political past, perpetuating the recursive world-building practices that foster human community by giving nature's givens a visible place within the legible world.

Vico's analysis helps us to see that scientists who refuse to participate in the politics of world-building may compromise their ability to speak for the things they study by undermining the political conditions that render nature perceptible and intelligible. He sees symptoms of an authority deficit in the axiomatic, adamant style of speaking of some of his Cartesian contemporaries: "When someone claims proudly . . . 'This is an axiom'; 'This is demonstrated,' he seems to me to be like a painter who writes beneath shapeless images that could never be made out on their own, 'This is a man'; 'This is a satyr'; 'This is a lion'; 'This is something else'" (*AW* 181). The objects for which this unduly proud scientist would speak, like the shapeless paintings comically labeled by their inept creator, are not *vera*: they are not "trues," but unintelligible blurs. To become intelligible and articulable, they would need to be *given* shape by a human maker.

The authority deficit of the hapless scientist in Vico's example—the reason her "talk has none of the force of truth upon the mind"—can only be partially understood by attending to epistemological problems with the validity of her demonstrations. This scientist has undercut the "compulsive force" of her words not merely by erring or lying, but by abstaining from the activities that reshape nature into a perceptible, speakable configuration—the world-building practices that accord to lions, thunderstorms, and other material phenomena their status as *vera* in the first place. Like the insufficiently creative painter, this scientist's axiomatic words cannot solve—indeed, they underscore—the problem of the inscrutability of the objects she seeks to represent. Vico considers the compensatory language of the Cartesian scientist to be "a lot of noise"—a clamor that signals a crisis of authority with deep costs for both science and politics.

Once again, Vico evokes the example of Descartes to illustrate the threat: "That only his judgment must be employed and only the geometrical method—that is too much. . . . Men will realize, too late however, that Descartes has done what those who become tyrants have always been wont to do" (*AW* 184). Vico accuses Descartes of compensating for his failed authority by accepting and reinforcing a shared condition of weakness, a classic move of tyrants. Science becomes an enemy to politics, from this Vichian perspective, when it seeks to suppress common sense, the power through which the world of nations is created. Vico's argument that common sense is a common root nourishing science and politics helps us to see that, when methods that devalue or disdain common sense stake an exclusive claim to the study of nature, the "world of nations" is threatened—a threat that is unwittingly courted by the two-sciences settlement. Like Arendt, Vico helps us to identify oppositional views of nature and politics as supports that undergird tyrannical versions of scientific authority.

But unlike Arendt, Vico does not consider antagonism toward common sense and the common world to be the defining or inevitable feature of the sciences. In *On the Most Ancient Wisdom*, he posits that the sciences "spring" from the same political contexts where good speakers sharpen their *ingenium*; the *New Science* extends the thought by arguing that "the sciences . . . sprang from the nations and from no other source." His writings affirm that science, at its origin, is a continuation of world-building politics by other means, which is precisely why he thinks it is so important to resist the perhaps inevitable fraying of science's ties to politics. Vico helps to open a critical vantage point from which we may distinguish between sciences that stake their authority upon augmenting versus dismantling common sense. He presses us to imagine what a modern corollary to Socrates might look like: a science trained in the public school of politics, conversant in both vernacular and erudite languages, and sharp enough both to recognize metaphorical connections forged in the past and to participate in creating new ones. Such a science would stand to contribute to the renewal of politics by fostering greater intimacy with the givens of nature.

If this is difficult to imagine, Vico offers his *New Science* as an example. He tells us he has sought to uphold "this criterion of truth: that we must respect the common judgment of men—the common sense of mankind" (*NS* 1406). As in the case of Socrates in Athens or the experimental culture Vico admired in Naples, when Vico engages the disparate particulars of the political past he seeks to reflectively re-create things that were originally rendered intelligible by the unreflective, shared judgments of common sense. From an Arendtian perspective, such an approach to the science of nature, as a *part* of the science of politics, might still seem to risk repeating the mistakes of Descartes: Vico's ethos of creative worldly engagement stands to make it that much more difficult for man "to experience the reality of what he himself is not."[19] But I hope my reading of Vico has shown that the radical creativity he attributes to common sense is more receptive to the givenness of nature than Arendt assumed. In addition, Vico's *New Science* promises to foster intimacy with another source of otherness cherished by Arendt: the past. His imaginative excavation of the earliest founding moments of human history recovers the otherness of ancestors who are barely recognizable as such. Vico affirms the memory of these past acts of world-building as an integral part of the act of knowing on the part of scientists today.

Practicing what he preaches, Vico establishes novel relationships between human history and nonhuman nature in the *New Science* in and through the relationships he establishes between the beginning and the end of history—as tremendous a feat of *ingenium* as ever there was. As Fisch captures so well, Vico's "new science comes . . . with a *re*making, a *re*constructing. . . . The remaking not only represents but continues and is part of the first making, and partakes of its reality."[20] Vico's new science partakes of the reality that only common sense can create and knows it in the only way he thinks is available to the human creature: by adding his name to the multiauthor, iterative work that is the world. He derives authority from his participation in authoring the world for which he speaks.

CHAPTER 3

Descartes and Democracy

It is difficult to think of another philosopher of René Descartes's stature who had less to say about politics. Yet generations of readers have seen political significance in Descartes's very turn inward, away from the political realm, to think in earnest about thinking. Important figures in the French Enlightenment viewed Descartes's epiphantic *cogito, ergo sum*—"I think, therefore I am"—as heralding the dawn of a new, more democratic era.[1] For example, Jean le Rond d'Alembert eulogizes Descartes as a "leader of conspirators," who showed "intelligent minds how to throw off the yoke of scholasticism, of opinion, of authority . . . preparing a resounding revolution" for "a more just and happier government."[2] In stark contrast, recent political thinkers more often view the Cartesian *cogito* as an emblem of elitism, complacency, androcentrism, imperialism, and other forms of domination.[3] For example, Sheldon Wolin argues that scientific reason in the Cartesian vein "is tempted into a certain politics that entails discrimination, exclusion, and simplification" and supports a "version of 'reason of state.'"[4] As we have seen, Arendt participated in and Vico helped to originate this line of criticism. Vico and Arendt each faulted Descartes for devaluing common sense and pluralistic world-building politics in favor of the worldless privacy of solitary contemplation and mathematical abstraction. Today, Descartes's pervasive reputation as a founding figure of modern scientific rationalism has made him a familiar villain in political theory's narratives about the early modern period and its influence on contemporary politics.

Rather than choosing between the democratic and the despotic Descartes, this chapter complicates his positioning as a thinker of thinking in these divergent appraisals of the Cartesian legacy. Descartes's recent critics, like his Enlightenment admirers, privilege one Cartesian moment at the expense of another: they recount Descartes's meditative seclusion in his legendary stove-heated room, but neglect his emergence to recount what transpired during his solitude, making a legend—or as he calls it, a fable—of his private thoughts. Where Descartes's thinking about thinking draws inward, his writing about thinking turns outward. In the *Discourse on Method* (hereafter cited parenthetically as *D*) as well as an earlier work of physics, *The World* (hereafter cited parenthetically as *W*), Descartes addresses readers in the vernacular French and adopts the folk genre of the fable. He makes a public spectacle of the private, invisible, and (in Descartes's characterization) immaterial activity of thinking. Through close readings of these texts, I show Descartes to be at once a solipsist and a publicist, a loner and a founder, a philosopher and a poet. Taking both sides of his writing about thinking seriously and attending to the role of aphorism, dialogue, and fable in his corpus will show his "new science" to have democratic leanings overlooked in familiar Enlightenment affirmations as well as democratic costs overlooked in familiar contemporary critiques.

I develop this alternative perspective on Descartes by adopting aspects of the genealogical approaches to the politics of science that I recovered from Arendt and Vico in earlier chapters. Like Arendt and Vico, my central question about the rise of this—or any—new science is not whether Descartes knows what he claims to know, but how he involves himself in the world-building practices that determine the conditions under which material phenomena appear and may be perceived and spoken for. To recognize traces in Descartes's writings of his participation in world-building politics requires keeping one's head tilted at an odd angle, as it were, to resist Descartes's own persistent efforts to center considerations of truth and validity. Arendt and Vico each model the right posture, but they straighten their necks too soon and thus come away with only half the story: the part about Descartes eschewing common sense and abandoning the common world in the name of guaranteeing the validity of his new science. Working with and against both Arendt and Vico, I show that Descartes first engages with common sense in order to create the worldly conditions for his own withdrawal and to engender public esteem for a privatized mode of scientific conduct. Descartes is, of course, keen to impress his readers with the indubitable validity of the *cogito, ergo sum* and the other truths that he claims to have derived from this first one. But Descartes also recognizes that authority does not follow inexorably from validity, which is why he needed to write more than a three-word proof to establish the foundations for his new science.

To lay these foundations, Descartes acts publicly to make his private acts of thinking appear, as if in a picture, and calls upon men of common sense to serve as spectators and judges of these spectacles of certainty. In Arendtian terms, he uses his writing to open a space of appearances, within which he seeks to guarantee the "realness" of acts of thinking that would otherwise be invisible and unreal. Descartes enlists his readers to act as his collaborators in world-building practices that would install the new science as a visible and authoritative presence just beyond the border of the common world and the reach of the common man. In these respects, I argue, Descartes's fables are "politics" in Jacques Rancière's sense of the term: Descartes writes to interrupt an entrenched "partition of the perceptible" by audaciously entrusting the task of judgment to a segment of society whose members lack any acknowledged authority to speak about the world they inhabit.[5] But if Descartes the fabulist animates an unconventional audience on behalf of the new science, he also seeks to discipline it—to open *and* to close politics in Rancière's sense. My readings show Descartes to be an early exemplar of modern science's ambivalent wavering between publicity and privacy, democratic openness and despotic closure, and realness and truth.

I find in Descartes's fables, aphorisms, and dialogues, then, not his "missing" political philosophy, but overlooked instances of political *action*, in which the shape of the world would be reworked in and through the reorganization of the "we" who inhabit, observe, judge, and speak for it. These fables are "rhetorical" in the sense that Vico's account of *ingenium* taught us to appreciate: Descartes deploys figurative speech not to manipulate "tame audiences" by dressing the truth in "conceptual vestments," as John Schuster has it, but to create worldly, embodied relationships among men and women, in the plural, and the material things they encounter on their diverse paths through the world.[6] Descartes seeks not to hide or distort nature with his rhetorical performances, but to reconstitute nature's givens and their human spectators into a world of relationships. He would orchestrate a new organization of feeling, habit, and comportment among the members of his vernacular reading audience, calling them to desire and actively participate in creating a world in which Descartes alone would have the standing to act as a spokesperson for nature. This ambivalent mutuality between the authority of the people to judge and speak for science and the authority of the scientist to judge and speak for nature may still be an important facet of what we refer to as "science" today, and an important reason why the relationship of science to politics remains important and confounding in democratic theory and practice. However foreign Descartes's *method* may be to today's researchers, and however dubious his supposedly certain proofs may be to contemporary philosophers, his fables are illuminating precursors to contemporary struggles over the authority of science in our democracies and our worlds.

Everyman Against the Philosophers

> I despaired of learning any truth about these things while I was on earth, and I thought the only escape from all my difficulties would be somehow to acquire wings and go up to heaven. My eagerness chiefly gave me this hope, and also the fabulist Aesop, who opens up heaven to eagles and beetles, and sometimes even to camels.
>
> —*Lucian of Samosata*[7]

The World is a treatise on physics, written by Descartes between 1629 and 1633 and unpublished in his lifetime.[8] Therein, he propounds a mechanistic cosmology and rejects the Aristotelean principles of substantial forms and final causes canonized in Scholastic natural philosophy. He redescribes the various visible and invisible manifestations of matter in terms of the speed and density of moving elemental particles. At first, Descartes adopts the didactic tone and deductive techniques that his reputation as a rationalist prepares us to expect: he exposes what he considers to be the dubious assumptions, misleading empirical evidence, and "subtle" reasoning of Scholastic natural philosophy and touts the clarity and distinctness of his alternative explanations. It is simply not possible to disagree with his arguments, Descartes suggests, because they are correct. For example, Descartes instructs you to "examine as much as you please all the forms that the diverse motions, the diverse shapes and sizes, and the different arrangement of the parts of matter can lend to mixed bodies," and takes your assent to his view to be inevitable: "I am sure you will find none" that defies atomistic explanation (*W* 42–43). In Descartes's characterization, his arguments will compel you to believe because they are true, not because "I want to force you to believe" (*W* 37). The authority of his physics, he suggests at many points, will follow from its validity.

But after five brief chapters in this vein, Descartes interrupts himself to introduce a surprising genre change: "Many other things remain for me to explain here, and I would myself be happy to add here several arguments to make my opinions more plausible. In order, however, to make the length of this discourse less boring for you, I want to wrap part of it in the invention [*l'invention*] of a fable,[9] in the course of which I hope that the truth will not fail to come out sufficiently and that it will be no less agreeable to see than if I were to set it forth wholly naked" (*W* 49). In a move that seems superfluous in light of the trust he already expressed in the power of reason to compel agreement, Descartes proposes to supplement argument with fable. What is the imaginative device of fable doing in a scientific study of the principles of matter and motion? What does it suggest about Descartes's ambitions as a physicist that he attributes to fable a productive role in making the truth "come out" in *The World*? And how should it impact our assessment of the democratic and scientific significance of Descartes's legacy to notice his recourse to fable, a didactic and fictional genre long considered to be of and for the people?[10]

Descartes explains that he turns to fable out of concern that he might lose "you," his imagined reader, to boredom if he were to sustain an unadulterated rational mode of address. He avows that he would be happy enough to continue presenting his arguments "wholly naked," adding nothing to supplement their inherent agreeableness. But he also anticipates that not everyone worth reaching will share his taste for these truths. The metaphor through which he introduces his fable acknowledges that truth, like the human body, can be made even more pleasing by dressing it up—not least because clothing reminds us of the nudity it teasingly deprives us of seeing.[11] By wrapping his discourse in fable, Descartes complicates his previous expressions of confidence in the inherent attraction of arguments that are true. By making his treatise more pleasurable, he seeks to tantalize readers in a way that truth alone, no matter how plausible, could not. He trusts that his appeals to their passions will not hide the truth, but help it to appear.

Descartes's turn to fable is an inviting, inclusive gesture—a measure to popularize the science of *The World* by adding something akin to sex appeal to its arguments. To observe this much adds some credence to the Enlightenment thesis of the democratic Descartes, while also raising questions about the kind of democratic spirit he would conjure. Descartes does not conform neatly to his Enlightenment interpreters' assumption that science appeals to a people defined by their equal natural endowment of reason. He appeals not only to his readers' reason, but also to their tastes and passions. Insofar as their assent is a matter of taste, he cannot count on it being given inevitably or universally. Taste, unlike reason, may be tempted, but cannot be compelled. In this sense, "the people" to whom Descartes appeals are not an established fact, fashioned by nature in bestowing reason equally to all. Descartes writes this treatise as a fable to captivate and thereby to constitute an audience for science that does not yet exist. He wraps his arguments in the invention of fable in order to invent the people whom he would have receive them.

This turn to fable at the same time risks alienating another, more likely audience: the Aristotelian Scholastics, who share Descartes's love for argument even as they disagree with his opinions.[12] Descartes does not avoid this risk—he courts it. *The World* at once offers to "you" the pleasures of fable and withholds from "them" the pleasures of debate. Rather than trying to convince Aristotle's followers of the implausibility of their views, Descartes says, "I am willing that they think themselves correct, for I have no intention of stopping to contradict them" (*W* 57). Though he thinks that Scholastic doctrines are false, Descartes will not pause to right this wrong in *The World*. Instead, taking a page from the fabulist Lucian, he intends to escape from the debates of the philosophers by imagining a flight into the distant reaches of space, where he will be free to envision the birth of a new, fictive world composed of vibrating particles of matter. "My plan is not to set out (as they do) the things that are in fact in the true world, but only to make up as I please from this matter a world in which there is nothing that the densest minds are not capable of conceiving, and which nevertheless could be created exactly the way I have made it up" (*W* 57). Fable permits Descartes to simply abandon, without refuting, the established wisdom and its defenders, giving himself—and "you"—a fresh start in a new, unknown, but easily conceived world. The turn to fable is both inclusive and exclusive, inviting even the densest of readers to follow Descartes in leaving behind the learned philosophers and become the newly privileged observers of a new, made-up world.

The gaze of his imagined readership provides Descartes the fable writer with a crucial enabling constraint: lest his imagined alternative to the "true world" of the Scholastics be mistaken for a *mere* fiction, he intends to ensure that anyone, no matter how dense, could conceive it along with him. In the name of science, he depends on a kind of common sense—what anyone could conceive—to set limits as he explores the pleasures of invention.

Descartes's appeals to the easy conceivability of his fable echo but also differ from his earlier appeals to the universal agreeability of his arguments. There is no compulsory answer to the question of what anyone could or could not conceive about a fictional world; nor is there anything clear or distinct about the properties of a world that does not yet exist. Descartes's appeal to common sense qua easy conceivability is not an appeal to the self-evident or the conventional; on the contrary, it is a plea to escape from the established world picture and imagine familiar things in a new way. To rise to this self-imposed standard, Descartes's fable must, in the tradition of Lucian and Aesop, reconnect the quotidian with the transcendent, the obvious with the unprecedented, and the senses with the imagination.

By holding Scholastic philosophers to this standard, Descartes shows that they have committed not only the epistemological errors of arguing falsely or illogically but, moreover, the political error of disregarding common sense. Descartes faults them for shoring up their long-standing, but deteriorating, authority to speak for nature by deflecting the scrutiny of the wider audience he would involve. Scholastics "themselves avow that the nature of their motion is very little known. To render it in some way intelligible, they have still not been able to explain it more clearly than in these terms: *motus est actus entis in potentia, prout in potentia est*, which terms are for me so obscure that I am constrained to leave them here in their language, because I cannot interpret them" (*W* 63). In a conspiracy of form and content, Scholastics wrap their "difficult to conceive" notions of matter and motion in an obscure Latin vocabulary, protecting them against outsider oversight (*W* 57). Descartes instigates a counter-conspiracy, writing in French and wrapping his "easy to conceive" arguments in the invention of a fable. He invites the outsiders in to scrutinize his new science and in so doing makes visible the terms of their exclusion from the established organization of the activity of knowing. The wrong that *The World* aims to right concerns, first and foremost, not the invalidity of opinions accepted as true by an expert elite, but rather the construction of their expertise as a walled-off echo chamber. To repurpose Rancière's terms, Descartes critiques the "partition of the perceptible" of the Schools for assigning the authority to speak for nature exclusively to those who speak most inscrutably and in a dying tongue.

Descartes similarly dramatizes this wrong and its overcoming in *The Search for Truth by Means of the Natural Light*, a dialogue occasioned when

Epistemon (a Scholastic philosopher) and Polyander ("Everyman," "the person of untutored common sense") together ask Eudoxus (Descartes's stand-in) to teach them everything he knows.[13] Epistemon, the Scholastic, soon finds he cannot tolerate the role of student and angrily claims the part of teacher: "I am going to show you what Logic can do when roused," he threatens![14] Eudoxus, the Cartesian, surmises that Epistemon's ostensible desire to know the truth is a cover for his deeper desire for—and fear of—domination: he "puts forward his own authority as reason and is anxious that others should submit to him in the way that he himself once submitted to others." Eudoxus refuses to continue a conversation in which authority would be won from submission, and demands a further changing of roles. He will go forward with the search for truth "only on condition that you," Polyander, stop playing the part of student and begin to "act as judge in our dispute, for I dare not hope that Epistemon will give in to my arguments. . . . But I shall have cause to be content . . . if only you agree with what I shall say."[15] Eudoxus's ultimatum, like the introduction of fable into *The World*, suspends a debate about the validity of competing representations of nature in order to initiate a dispute over who may legitimately author and judge these representations. As in *The World*, Descartes temporarily pushes both Epistemon and epistemology into the background and prioritizes the creation of a mutually contenting solidarity between his new science and the everyman.

Rancière can help us to see in Descartes's efforts to foment a dispute concerning "the existence of parties as parties" a kind of political significance that debates about the democratic Descartes overlook.[16] Rancière would "reserve the term *politics* for . . . whatever breaks with the tangible configuration whereby parties and parts or lack of them are defined by a presupposition that, by definition, has no place in that configuration—that of the part who have no part."[17] In politics, according to Rancière, those who are considered to be of no account in a debate suddenly intrude uninvited, and their act of defiance renders newly visible the wrongness of their hitherto naturalized exclusion. Descartes's fable is politics in Rancière's sense: it interrupts an argument with Scholastics to beckon to readers whose institutionalized irrelevance to philosophical argumentation had served as an unnoticed guarantee of the coherence of Scholastic authority. Inviting "you"—the undistinguished reader of vernacular French, whom Descartes hails repeatedly—to take the part of judge, Descartes would repartition the perceptible. His fable takes the first steps toward revolutionizing the sciences of nature by opening the way for "everyman" to take part in a conversation about the nature of matter and motion.

The World of Our Imagination

Descartes's next step is to invite this partless part to join him in leaving behind the old world and witnessing the creation of a new one: "For a short time then, allow your thought to wander beyond this world to view another, wholly new one, which I shall cause to unfold before it in imaginary spaces," where "we can lose from view all the creatures that God made five or six thousand years ago" (*W* 49–50). As Descartes beckons to you to enter imaginary spaces, he shifts tense, interpolating "you" into an imagined "we." As the old world disappears, and before the new one appears, new relationships unfold in the imaginary spaces between the author and his readers. As creator of the fable, Descartes invites this unfolding, but he cannot, alone, cause it. He stages an event akin to what Jason Frank calls a "constitutive moment": he seeks "to elicit and enact the very people on whose authority [his] claims are made," drawing "authority from the future anterior, from the political horizon of what will have been."[18] The "we" of *The World* and the authority of its author are (yet) to be born in the imagination of its readers. We conceive ourselves as a collective if and when we together accept Descartes's invitation to imagine the disappearance of the old world. In both of the double meanings of "conceive," we at once comprehend and create ourselves as fellow wanderers, united by a collectively imagined view of nothing from a collectively imagined view from nowhere.

Thus far, Descartes has maintained his distance from this emergent collective, reminding us of his singular role—"*I* shall cause"—in conjuring the appearance of the new world before us. But he soon closes this distance: "Suppose that God creates from anew so much more matter all about us that, in whatever direction our imagination can extend itself, it no longer perceives any place that is empty" (*W* 51). Descartes's narration of the moment of world-creation at once submerges his authorial "I" into "our imagination" and names God as the supposed author of the new world. He doubly displaces authorship from himself to two higher powers: not Descartes, but God shall create the world's matter; and not Descartes's singular vision, but "our imagination" shall create this picture of divine creation. The first collective endeavor of this nascent "we," after leaving behind the true world and its true God, is to imagine a new God, giving him the power to give us an imaginary plenitude. In this sense, even the authorial role that Descartes displaces onto the imagined God redounds upon the imagining "we": the power to create a new world, and even to create its creator, is ours, not His (God's) or his (Descartes's). As "our imagination" extends itself, it fills every empty place with matter, an autopoietic transubstantiation of image into substance. Descartes shares in this collective creative power, but only insofar as he continually subsumes himself within the collective imagination, living up to his professed responsibility to narrate a fable that anyone could conceive. If his narrative should wander too far, reaching imaginary spaces that are inaccessible to the everyman, he would be unable to produce anything as common or as extraordinary as a new world. The author of *The World* cannot author the world alone. The conception (in the sense of generation, invention, or inception) of the new world depends on its conception (in the sense of perception, picture, or idea) by his readers.

As Descartes distributes the power of world-constitution to this imagined, imagining readership, he also shares with his readers the obligation to create a world that is easy to conceive. "Since we are taking the liberty of imagining this matter to our fancy, let us attribute to it, if you will, a nature in which there is absolutely nothing that anyone cannot know as perfectly as possible" (*W* 38). The "liberty of imagining" comes with the responsibility to see the world through the eyes of innumerable others, anticipating what they could know and accepting this anticipation as a governing principle for "our fancy." To anticipate how an imagined world might appear to others itself requires an act of imagination—a point that Arendt emphasizes in her account of "enlarged thinking." She argues, evoking Kant, that thinking is enlarged when "by the force of imagination it makes the others present and thus moves in a space that is potentially public. . . . To think with an enlarged mentality means that one trains one's imagination to go visiting."[19] The imaginary spaces that Descartes beckons us toward are "potentially public" in Arendt's sense: a space to go visiting, seeing the world anew by imaginatively occupying various vantages upon it. The world of "our imagination" becomes a potentially public thing, a *res publica potentia*, through an exercise in enlarged thinking.

Reading Arendt against Arendt helps us to see that Descartes does not abandon the common world when he abandons the old world, for he strives to keep "our imagination" tethered to common sense throughout our journey into space. As we have seen, her concern about science in the spirit of Descartes is that, in the search for truth, he and other scientists too readily abandon common sense, endangering the very realness of the phenomena they seek to know.

The World places this interpretation of Descartes's legacy in question: in this text, Descartes refuses to leave Polyander behind. In the service of science, he enlists common sense, casting it not as a static repository of self-evident truths, but—as Vico does—as an active, creative, world-building faculty. The fable, as a tether between common sense and imagination, pulls at both ends, spurring the people to reimagine whatever seems given or obvious. As H. J. Blackham explains, "The reader of a fable has to know what he sees in the image presented and to see it in a new light."[20] In *The World*, common sense serves as the starting point for its own transfiguration, creating unprecedented new possibilities for the shape of both the world and the "we." Temporarily eschewing the question of truth, Descartes wraps his arguments in fable in a way that raises the question of realness: might this fictive world emerge from the spaces of imagination into the corporeal reality that appears between us? Might it become a real alternative to the true-but-inconceivable world of the Scholastics by virtue of the involvement of its plural authors and spectators?

The common sense invoked in *The World* is a different kind of democratic spirit than Descartes is sometimes admired for, and a facet of Descartes's science that has been unappreciated by many of his critics. *The World* at once raises questions about whose perspective on the world counts and whether or not they share a common world in the first place. It advances what Rancière calls "the *demonstration* proper to politics," a demonstration that "is always both argument and opening up the world where argument can be conceived and have an impact—argument about the very existence of such a world."[21] The success or failure of this political demonstration—and with it, Descartes's plan to refound science—hinges on a constitutive moment in which the "we" and the world would be reshaped side by side.

Descartes ultimately left the imagined audience and their imagined world of *The World* in abeyance when he chose to publish, in its place, the *Discourse on Method*. But we will see that he did not abandon the genre of the fable. Rather, he continued to experiment with new ways to animate "our imagination" in this second iteration of his new science, while also discovering new ways to discipline the people he enlisted in his cause.

A Feel for Truth

Still, common sense in the guise that we noticed it at work in *The World* is conspicuously absent from the portrait of mankind's natural intellectual endowments that opens the *Discourse on Method*. Instead, Descartes's first lines center on a cognate term that means something quite different than common sense as Arendt, Vico, or I use the term: namely, *bon sens*, or "good sense," as it is typically translated:

> Good sense is the best distributed thing in the world . . . even those who are the hardest to please in everything else do not usually desire more of it than they possess. In this it is unlikely that everyone is mistaken. It indicates rather that the power of judging well and of distinguishing the true from the false—which is what we properly call "good sense" or "reason"—is naturally equal in all men, and consequently . . . the diversity of our opinions does not arise because some of us are more reasonable than others but solely because we direct our thoughts along different paths. (*D* 111)[22]

Descartes equates "good sense" with reason, defines them both as the power to judge truth and falsity, and affirms that each man is endowed by nature with an equal share of this power. But "good sense" in this passage is more than a synonym for natural reason; it also names a specific, distributed way of organizing the act and the affect of reasoning. Men of reason live as men of good sense when each is content to judge alone, from separate but equal paths.[23] Good sense aligns even those who are hardest to please around the principle, to each their own. Adherence to this ethos is underwritten by a peculiarly solitary, self-satisfied temperament, free of envy or curiosity concerning the opinions formed by other men walking other paths. Men of good sense, as Descartes portrays them, have little in common other than their mutual lack of desire for shared truths.

The ethos, temperament, and habits of good sense, thus defined, are strikingly at odds with *common* sense, that collective, creative, public mode of judging that I identified as integral to Descartes's project in *The World*. Indeed, a person shows good sense precisely by abstaining from anticipating how others might conceive and judge the world. In Arendt's vocabulary, men of good sense eschew enlarged thinking, preferring the contented repose of thinking narrowly from where they stand. In short, the "good sense" evoked by Descartes is a radically depoliticized, solipsistic comportment toward the world—a way of seeking truth alone that Arendt would rightly flag as a risk to the very realness of the world. But what Arendt does not recognize, and precious few of Descartes's readers appreciate, is the extent to which his efforts in the *Discourse* to refound science involve him in a struggle *against* good sense, and thus against the very solipsistic tendencies that he is widely seen to foster. Descartes endeavors to overturn the intellectual habits, bodily comportments, and affective leanings that confine seekers of truth to separate paths; he seeks to awaken desires for shared judgments and commitments that are repressed by good sense.

However, if the Descartes of the *Discourse* is not the prophet of solipsism that his critics fear, he is certainly not the prophet of equality portrayed and praised by theorists of the democratic Descartes. Descartes's antagonism toward good sense is part of his broader opposition to equality in the *Discourse*—a marked departure from *The World*'s opposition to the intellectual inequalities of Scholasticism. Descartes stirs desires for common truths among men of good sense in the service of preventing them from claiming the authority to judge and speak to matters of truth and falsity, assigning this role exclusively to himself. He discovers a novel role for common sense in constituting the requisite worldly conditions to reorganize the search for truth around a principle of inequality. The new science of the *Discourse* depends not only on a new method, but also on a novel partition of the perceptible—a new way of seeing and of configuring bodies enacted in concert by the very people who are to be excluded from the search for truth.

Descartes's proof of the ubiquity of good sense in the *Discourse*'s opening lines is the opening salvo in a sustained effort to loosen the influence of good sense on the practices of his readers. This proof takes the form of an enthymeme, a probable conclusion drawn on the basis of ordinary experience. Descartes posits that anyone who takes a look around will notice what he considers to be obvious: namely, that most people seem satisfied with the opinions they have formed using the reasoning power at their disposal. On that basis, anyone could conclude, with Descartes, that reason is probably equally distributed; if it were otherwise, this contented state of intellectual peace would be disrupted more often by jealous contention. But in an era of good sense, the everyman has fallen out of the habit of looking around to take notice of other people judging from other paths. To follow the enthymematic reasoning that leads to the conclusion that all men are equally endowed with reason, men of good sense must first look up from the privacy of their paths; they must enlarge their thinking to anticipate the enjoyment that other judges must take in the judgments they have formed. The only way to arrive at the probable truth of rational human equality by way of Descartes's enthymeme is for his readers to violate the solipsistic ethos of good sense, animate their dormant liberty of imagining, and begin to exercise *common* sense.

But having invited his readers to see each other as equals in judgment, Descartes proceeds to portray himself as the exception to the rule of equality: "I consider myself very fortunate to have happened upon certain paths in my youth which led me to considerations and maxims from which I formed a method whereby, it seems to me, I can increase my knowledge gradually and raise it little by little to the highest point allowed by the mediocrity of my mind and the short duration of my life" (*D* 112). The famous pinnacle of this path is his discovery of the proof now colloquially abbreviated as "the *cogito*," though he writes it first in French: "*Je pense, donc je suis*," I am thinking, therefore I exist (*D* 127). Descartes says that he could not doubt this thought even after rejecting as false every other belief he previously held, including his faith in the reality of the appearing world. This is the moment Arendt has foremost in mind when she speaks of Descartes's hostility to appearances and Vico evokes when he compares Cartesians who noisily exclaim their demonstrations of truth to inept painters who must label their paintings.

Arendt's and Vico's criticisms are important but incomplete, neglecting the multiple appearances of appearances and visions of vision in the *Discourse*. From the outset, Descartes declares his intention to make a spectacle of his acts of thinking, and explicitly models the text upon the creation of a painting: "I shall be glad . . . to show [*faire voir*] in this discourse what paths I have followed, and to represent my life in it as if in a picture, so that everyone may judge it for himself."[24] He promises that the spectacle of his unusually ascendant path through life toward truth will be a sight worth seeing, tempting the eyes of readers accustomed to the blinkers of good sense. To follow this story is to follow Descartes far from the familiar terrain of good sense into imaginary spaces: the life of the mind of another.

Yet we should not mistake Descartes's recounting of his life story for a lesson in how to live or an invitation to imitate his exercise in radical doubt and its overcoming: "My present aim, then, is not to teach the method which everyone must follow in order to try to direct his reason correctly, but only to show how I have tried to direct my own. . . . I am presenting this work only as a history or, if you prefer, a fable" (*D* 112).[25] Descartes's autobiography is not a guidebook, but a fable. This is a poetic and painterly work, intended to woo men of good sense to imagine their way beyond the confines of their own minds and to judge what Descartes has done . . . but without doing it for themselves. Even as Descartes beckons to his readers to follow his narrative, he refrains from showing them how to follow his method. In asking us to judge a picture of his path, he makes strong demands on our imagination without directing our reason.

So while it is important to Descartes that we empathetically envision what it must have been like for him to feel such exceptionally unshakable conviction about the existence of his thinking "I," he neither teaches us how to become equally certain about our own thinking nor expects us to confirm the validity of the proposition *je pense, donc je suis*. Indeed, he gives us reason to doubt that the feeling of radical clarity—the feeling that convinced him of the existence of his thinking—*could* be taught, even if he wanted to teach it (which he tells us he does not want to do):[26] "There is nothing at all in the proposition '*I am thinking, therefore I exist*' to assure me that I am speaking the truth except that I see very clearly that in order to think it is necessary to exist" (*D* 127). This certainty, as Descartes characterizes it, is an entirely private experience, self-founded and self-sustaining, requiring no assurance from other judges standing on other paths. Descartes has no assurances to give to those who don't see what he sees, don't see it as clearly, or don't see clearly the difference between clear and dim vision.[27] The *cogito* is a kind of truth that cannot be taught by transcribing a syllogism into a book, for the proposition "*I am thinking, therefore I exist*" is not the source of the assurance.

Instead, Descartes steps forward to personify certainty without teaching it; he narrates the *cogito* moment so that anyone could imagine it as an event, but he does not demonstrate it so that anyone could know it as a truth. The *cogito* is a fable wrapped in the invention of a syllogism, a way of making Descartes's thinking appear to others without necessarily proving anything to them. Through this device, Descartes paints himself as a fount of knowledge, a figure of science whose thoughts are at once indubitable and incommunicable. *This* man of reason is a hyperbolic reincarnation of the man of good sense: blind to anything beyond his own existence and committed to disregarding what other people on other paths might think of his thinking. Unlike the enthymeme that opens the *Discourse*, the syllogism at its climax portrays an extraordinary experience that is not for everyone to have. The *cogito* is for us to picture, but for Descartes alone to think.

Descartes's critics thus have reason to see in this picture a politically problematic affirmation of radical solipsism, even misanthropy. But if Descartes presents himself as indifferent to our rational judgments concerning the truth of the *cogito*, he expresses deep investment in our aesthetic judgments concerning the appearance of the *cogito*. He reports feeling "extremely satisfied" with the "fruits" of his private reflections—a passion characteristic of good sense, as we have seen (*D* 112). But he lacks the good sense to be satisfied with this satisfaction. The fact that other spectators judge questions of truth and falsity differently than he does is to be expected, from Descartes's perspective. But the fact that others are just as pleased with the fruits of their own private reflections irks Descartes enormously. He entertains unpleasant doubts not about the validity of the truths he holds dear, but about their desirability: "I may be wrong: perhaps what I take for gold and diamonds is nothing but a bit of copper and glass." Descartes can only really enjoy the fruits of his method if we, too, judge them to be treasures, whether or not we judge them to be truths. He craves a feeling *about* his feeling of certainty that he cannot attain alone, for it requires others to feel attracted to that which he alone possesses, much as the enjoyment of owning gold and diamonds consists in part in the pleasure of seeing the glint of desire in the eyes of onlookers. In making a spectacle of his thinking, Descartes makes visible, too, his desire for his thinking to be desired, an unusual—even unseemly—deviation from the prevailing norm of self-satisfied good sense. We, his spectators, may transform Descartes, through the force of our shared but ultimately futile desire to know what he knows, from a frustrated deviant into a fully satisfied expert.

Descartes's new science is founded not only on his private feeling of certainty, but also on the formation of a people drawn together by their mutual desire for a truth they cannot touch. When Descartes wants us to want his thinking, he is craving something akin to realness as Arendt defines it, the feeling that the things I sense are likewise tangibly present for other spectators, even if their experience of them is quite different. As Arendt cautions, this feeling is all too easily lost in moments of private reflection and cannot be regained through proofs, no matter how certain: "It never occurred to him [Descartes] that no *cogitatio* and no *cogito me cogitare* . . . would ever have been able to convince him of his own reality . . . without fellow-creatures to assure him that what he perceived was perceived by them too."[28] But Descartes is not ignoring the fact that thinking does not ordinarily appear to others; he is seeking to overcome this fact by making his thinking newly visible in a fable. The *cogito*, among other moments in Descartes's fabulous intellectual autobiography, is a way of worlding his private thinking acts, transubstantiating the imperceptible, immaterial thinking substance, *res cogitans*, into a perceptible, tangible, desirable public thing, *res publica*.

The Pleasure Is to Be Mutual

But if thinking can be done alone, appearing requires the presence and receptivity of an audience of other people. I said before that the author of *The World* cannot author the world alone, and something similar should be acknowledged about authorship and authority in the *Discourse*: Descartes on his own lacks the ability to extrude his thinking into a space of appearances and remake it into a real fixture of the common world. The conception (in the sense of generation, invention, or inception) of the *cogito* as a newly perceptible, worldly being depends on its conception (in the sense of perception, imagination, or mental picture) by his readers. Granted, our conception of Descartes's thinking may be less than perfectly clear or distinct, but our acts of conceiving it, however dimly or densely, give it the realness that it would otherwise lack. Exercising our "liberty of imagining," the liberty named in *The World* but beckoned, too, in the fable of the *Discourse*, we act as midwives to Descartes's rebirth as the man of reason, thinking incarnate and spokesperson for the world.

To accomplish this transformation, men of good sense must shed their blinkered incuriosity and find pleasure in the certainty painted and personified by Descartes. As sure as he is about the truth of the *cogito*, Descartes is not sure we will like what we see when we catch a glimpse of his acts of thinking; he fears they "are perhaps too metaphysical and uncommon for everyone's taste," a concern that made him unsure of "whether I should tell you" about them at all (*D* 126). Yet Descartes determined that he was "in a way obliged to speak of" the *cogito* moment. His way of speaking, a fable of certainty, attempts to teach listeners to cultivate a taste for the method he will not teach them to follow. Descartes makes a spectacle of his uncommon thoughts to intervene directly at the level of affect and imagination. He stirs new desires and offers new pleasures to contend with the good-sense satisfaction of judging alone. His discoveries stand to become more like gold and diamonds the more we lose our taste for the fruits of our private judgments of truth and come to covet the fruits of his method. To see Descartes's thinking, as if in a picture, is to experience certainty not as a subjective compulsion of the mind, but as an object of bodily desire—an object located within our sights, but beyond our reach.

This feeling of desire may be shared in a way that Descartes did not think the feeling of certainty could. Certainty, as Descartes portrays it, is a feeling of clarity or distinctness concerning one's thoughts that is neither bolstered nor derogated by considering the thoughts of others. As we have seen, he only came to feel certain of his thinking by first closing his mind's eye to the presence of other judges standing on other paths. By contrast, in order to judge the *cogito* to be a treasure, I must first enlarge my thinking, imaginatively anticipating that other judges will likewise desire the certitude embodied by Descartes. To covet Descartes's certainty the way I would covet his gold or diamonds is to feel that I am not the only one who feels longing when I look his way; it is, furthermore, to picture how others would look at me with hunger in their eyes if the treasure were mine.

The feeling of coveting something attractive is, in this respect, not just a feeling about the thing, but a feeling about other people and about how they feel about the thing. John Locke has something like this in mind when he attributes the origins of money, and the economic inequality that money facilitates, to a tacit "mutual consent."[29] For gold to become a currency, it is not enough for a single eccentric to want to "give his nuts for a piece of metal, pleased with its colour," Locke tells us. For gold to be worth accumulating, the pleasure must be mutual. Moreover, this mutuality must be apparent to the members of the protocommunity whom it thereby binds together. In different ways, Descartes and Locke both suggest that the birth of a shared longing is the birth of a people, whether or not these affective bonds are ultimately formalized in a social contract. Insofar as Descartes's fable moves us to feel shared passions for shared truths, we will have transformed ourselves from men separated by our good sense into men connected by our common sense—connected by a feeling that our feelings are shared. Those who feel an aesthetic attraction to the spectacle of Descartes's thinking "I" act as midwives, too, to their own rebirth as a "we."

But Descartes, unlike Locke, aims to dispel as "mere" fiction the notion that the everyman could amass his own stow of treasures if only he labored industriously. Descartes repeatedly warns us not to let our newfound attraction to shared truths mislead us into wanting to participate in their production: "If I am sufficiently pleased with my work to present you with this sample of it, this does not mean that I would advise anyone to imitate it" (*D* 118). Descartes would teach you, his imagined reader, to imitate not his work, but the pleasure he takes in its products. Quite unlike *The World*, the *Discourse* calls upon Polyander to be receptive to, not active within, the new science.

The peculiar balance between desire and restraint that Descartes seeks to cultivate in the audience of the *Discourse* is prefigured by one of his earliest aphorisms: "Science is like a woman: if she stays faithful to her husband she is respected; if she becomes common property she grows to be despised."[30] In this simile, as in the *Discourse*, Descartes pictures a science that would have the public appeal of a private woman, admired by everyone for her very refusal to be enjoyed by just anyone.[31] The *Discourse* seeks to make science more like a woman by promoting an affective economy akin to marital monogamy, in which desire would turn to disgust if science were to become as well distributed or commonly possessed as good sense. The science of Descartes's imagination is a private mode of conduct that depends, for its public respectability, on the shared comportment and aligned passions of a reverent but remote people. To make a woman of science in this sense is to inscribe a new table of values within the bodies of men, in the plural.

Buying Time: A Social-Scientific Contract

The refounding of science in the likeness of a faithful woman would require both restraint on the part of the people and fidelity to the *vida methodica* on the part of the scientist. In the name of that refounding, Descartes encourages the formation of a "we" united by shared pleasure in science, but divided into producers versus spectators: those privy to the pleasures of discovering truth versus those who have learned to be satisfied with glimpses of truth exposed by those in the know. To foster this partition, the *Discourse* retroactively effaces the active role of "our imagination" in conceiving—in its double sense of generating *and* perceiving—the new world and the new science of *The World*. Descartes writes of the time spent working on this text, "My aim was to include in it [*The World*] everything I thought I knew about the nature of material things before I began to write it" (*D* 132). Descartes's narration retrospectively divides the moment of knowledge production from the moment of textual production, belying the close connection between the constitution of a new science and a new people in *The World*. He thereby divides the two meanings of "conceive," ascribing the generative half to himself alone, as though he had produced *The World* without first considering what would be conceivable to the everyman.

The other half of "conception"—perception—he ascribes to us, his imagined readers: "So that you might see how I dealt with this subject, I shall give my explanation of the movement of the heart and the arteries" (*D* 134). Readers who receive this portrayal of the heart get a glimpse of both the nonappearing contents of the human body and the nonappearing contents of the unpublished *World*. Descartes bares just enough of this internal organ to underscore his point that the body—like the rest of the world's "material things"—is for him to make visible and for us to see at his pleasure. Unlike *The World*, where he dressed science in an alluring fable in order to ensure that "the truth will not fail to come out," the *Discourse* exposes a tantalizing example of his physical science in order to prevent other pieces of his corpus from appearing. To teach his readers to enjoy science the way they enjoy chaste women—at a respectful distance—Descartes makes this "sample of my work" appear and attract, only to withdraw the rest to a place where it can be cherished for its inaccessibility. He presents the science of bodies, like the metaphysics of the mind, the *je pense, donc je suis*, as the preconceived product of his private labor, afterwards selectively curated for public consumption and enjoyment.

Descartes follows this retrospective elision of the productive role played by the people in the science of *The World* with a parallel move to prospectively prevent the audience of the *Discourse* from becoming involved in the ongoing search for truth. Descartes wants to finish alone what he (claims to have) started alone: "If there was ever a task which could not be accomplished so well by someone other than the person who began it, it is the one on which I am working" (*D* 148). He entertains but quickly rejects the possibility of collaboration on the grounds that "voluntary helpers, who might offer to help him from curiosity or a desire to learn, usually promise more than they achieve." Worse, they "would inevitably wish to be rewarded by having certain difficulties explained to them, or at any rate by compliments and useless conversation, which could not but waste a lot of his time." Curious volunteers would inevitably slow the progress of science, frustrating the desire to learn that motivated them to volunteer in the first place. He urges those who are "eager to help" to resist this temptation, and offers an alternative way to act on their passions: "I do not see how they could do anything for him except to contribute towards the expenses of the observations that he would need and, further, prevent unwelcome visitors from wasting his free time." Our work is to secure his work against interference by other overeager spectators, providing the security and funding that he needs to go it alone.[32] Descartes calls upon us to police ourselves as well as each other in order to secure an asymmetric relationship between us and him. He cannot secure his "free time" without our complicity.

With the freedom secured by our involvement in leaving him alone, Descartes promises to put his time to responsible use:

> I have resolved to devote the rest of my life to nothing other than trying to acquire some knowledge of nature from which we may derive rules in medicine which are more reliable than those we have had up till now. Moreover, my inclination makes me so strongly opposed to all other projects, and especially to those which can be useful to some persons only by harming others, that if circumstances forced me to engage in any such pursuit, I do not think I would be capable of succeeding in it. Of this I make here a public declaration. (*D* 151)

Descartes closes the *Discourse* by vowing publicly to labor privately for the common good. Advancing the science of medicine, he argues, will help everyone and hurt no one. A few pages earlier, he is even more emphatic about the supreme value of sciences that serve "the maintenance of health, which is undoubtedly the chief good and the foundation of all the other goods in this life" (143). He reminds us that we can only enjoy "the fruits of the earth and all the goods we find there" while we are alive—we cannot take our gold and diamonds with us when we die.

Time, Descartes argues, is "undoubtedly" the ultimate treasure because nothing else can be enjoyed without it. He proposes the outlines of an asymmetrical, biopolitical social contract, with time as the medium of currency: if we commit to giving Descartes the uninterrupted time he needs to discover the causes of sickness and health, he promises to give us the knowledge we need to forestall life's interruption by death. The terms of this contract would add the weight of obligation to the pull of the passions, formalizing a distinction between the people, who desire truth but may not seek it, and the scientist, who must pursue and present truth for the sake of public health.[33]

In the terms that I am adopting from Rancière, Descartes's sketch of this social-scientific contract is a proposal to repartition the perceptible, to create public conditions to ensure that the authority to speak for material bodies becomes the exclusive provenance of a few exceptional figures. Insofar as we endorse these terms, we formalize and further cement our status as a people, even as we formalize and entrench the status of the people as the part that has no part in producing science. Unlike *The World*, where Descartes called on the people to assert their judgment against the domination and inequality of Scholasticism, the *Discourse* calls us to instantiate inequality anew, breaking with tradition to authorize the new science against the old.

But Descartes cannot forge this break alone; the contract he proposes may be asymmetrical, but it is mutual. It gains coherence from our consent, whether explicit (a "public declaration" to match Descartes's) or tacit (day-to-day ways of seeing, feeling, and acting that casually express deference to science and desire for the life-enhancing truths it promises). Insofar as we follow Descartes's call and consent to revoke the right and responsibility to participate in seeking the truths that we desire, we make real the inequality that Descartes, alone, could only imagine.

Conclusion

This reading of the *Discourse on Method* calls into question its traditional reception as an early celebration of human equality and a harbinger of Enlightenment conceptions of democracy. On the contrary, in a reversal of his assertion of the principle of equality in *The World*, Descartes seeks to subvert equality in the *Discourse*. The Descartes of the *Discourse* encourages the adoption of an affective economy and a social-scientific contract that would distribute the right and responsibility to speak for nature unequally, reengineering the equal distribution of good sense.

In so doing, he unwittingly highlights the instability of his own claims that human reason is a natural given. The man of reason who emerges from the pages of the *Discourse*—the remote scientist, empowered to present life-preserving truths to an eager, receptive public—is the product of collective political enactments, not of nature. Read closely, the *Discourse* reveals the new science to be a political artifact—a way of accounting the speech of some as *logos* and discounting the speech of others as having to do with pleasure and pain, not truth and falsity. Reading Descartes against Descartes and with Rancière and Arendt helps us to see that Cartesian reason is not a natural essence, lurking unseen within the mind until the day it is awoken and trained through mental calisthenics. Rather, Cartesian reason becomes real and tangible, and is granted its potency *as* a distinctive, individuating possession of the thinking "I," by virtue of the ongoing passions and actions of the "we." Like the chastity of a woman, the man of reason gains coherence and realness through a politics of orchestrated feelings, habits, values, and comportments—a politics that may, with difficulty and struggle, be contested and reorchestrated. The power of reason on which Descartes stakes his new science is an accomplishment neither of nature nor of the private mental discipline of one René Descartes but of the people.

In this respect, it would also be wrong to follow recent currents in democratic theory in viewing Descartes's rejection of equality in the *Discourse* as a rejection of democracy tout court. Good sense, as he paints it, is a particularly shallow, bloodless, and atomistic version of equality. The spectacle of certainty presented by Descartes for our judgment disrupts at once the equality and the isolation of men of good sense. He calls them toward one another even as he pushes them away from himself and his science. Descartes enlists the people to subvert their own equality and, in so doing, fosters their formation *as* a people, united by what I, inspired by Arendt and Vico, have been calling common sense: a sense for the world that begins by sensing that it is shared. Descartes's opposition to good sense in the *Discourse* creates openings for the fellow-feeling of common sense, even as he enlists the people to close the door between common sense and science.

Whether we choose to throw our weight behind shutting the door between ourselves and the search for truth today has a lot to do with whether we share Descartes's unshakable belief that health is "undoubtedly the chief good of life." As Descartes has taught us, just because *he* cannot doubt a proposition does not mean that others will necessarily share his certainty. Descartes is not keen to "make visible" the prospect that other people standing on other paths might question the supreme value of health and, therewith, the obligation to protect the life-preserving sciences from public interference.[34] Yet such dissent is easy to anticipate—all the more so after reflecting on the *Discourse*'s diagnosis of the intractability of disagreement. For example, might some people sometimes place a higher value on courage, the willingness to sacrifice one's life for the sake of a larger good?[35] Might others doubt that the health benefits of the sciences could ever be so evenly distributed that we could say in honesty that they help everyone and harm no one? Descartes's assertion of the supreme value of health, meant to seal the people's agreement to the terms of the contract he has proposed (and cement their status *as* a people), provides an opening for further dissensus and future repartitioning of the perceptible.

If and when we do unite as a people to affirm the value of life-giving truth, despite these and other qualms, we do so not under the compulsion of an indubitable moral principle, but as a provisional consequence of democratic judgment. As Descartes himself points out, judgments of value never have the finality of perfect certainty, making it often "necessary to act upon opinions which one knows to be quite uncertain just as if they were indubitable" (*D* 126). When we act to preserve the autonomy of science—as many of us do, for example, in the face of the mortal danger of climate change and its deniers—we do not act under the compulsion of our certainty, nor that of the scientists. Rather, we do so on the basis of a provisional political judgment, informed by our common sense—our feeling when we look at each other and at the world around us that it is sometimes worthwhile, even courageous, to sacrifice equality for the sake of health, our own and the world's. Descartes thus helps us to see, if unintentionally, that deference to science, when it occurs, is not the automatic or obvious outcome of "mere" opinion bending its knee before truth, but one contingent possibility of democratic politics. Such deference is rooted in an enlarged way of judging value and beauty and the collective practices that reflect and reinforce these judgments. What might look like antidemocratic tendencies embedded in pro-science wings of the environmental movement look much more complicated when read in light of my reading of the *Discourse.*

But if the Descartes of the *Discourse* can help us to recognize democratic significance in calls for greater deference to science coming from center-left public health movements today, the Descartes of *The World* can help us to see the democratic significance of radical movements that contest the politics of expertise and lay claim to science as the work of the people. We can follow the *Discourse* in rejecting the isolation of good sense while refusing to relinquish the principle of equality in the search for truth. We can emulate Polyander and Eudoxus, raising suspicions about anyone—including Descartes and his heirs—who claims to have neatly divided the roles of student, teacher, and judge or to have inhabited one role to the exclusion of others. We can affirm the citizen scientists of ACT UP or Woburn, Massachusetts, by refusing to account speech as *logos* unless and until it reflects a sincere effort to take into account what various observers have seen from their disparate paths as patients, mothers, and neighbors. We can follow *The World* in insisting on being connected to each other as equals in and through the practice of anticipating and articulating differences of opinion about what the world is and what it calls us to do.

CHAPTER 4

Hobbes's Worldly Geometry of Politics

In preceding chapters, Arendt, Vico, and Descartes helped me to elaborate an unconventional account of "the world" as a constellation of beings—human and nonhuman, organic and inorganic, given and made—drawn together and related through concerted, embodied political practices. I have argued that, over the course of struggles to give shape to this relational reality, the world's various human and nonhuman parts are granted or deprived of visibility and legibility even as some human actors (and not others) come to be heard and acknowledged as speaking for those parts that cannot speak for themselves or are not heard as such. The politics of world-building, I have maintained, is the right context in which to consider the foundations of the authority of the natural sciences to speak for their objects of study. But this way of complicating traditional ontological oppositions between nature and politics also suggests the need to reconsider the basis of the authority exercised and contested by political scientists and political theorists, and by the human actors who are their objects of study. If every object of every science is a body politic, deriving its realness, perceptibility, and susceptibility to representation from its place in a politically constituted, heterogeneous world of relationships, then it is not possible to neatly separate the question of who speaks for nature from the question of who speaks for the people. Against the two-sciences settlement, the authority of the natural and political sciences—and the authority of sovereigns, citizens, and other human subjects—are contested and settled in tandem, as various actors struggle with and against each other to build a common world and to decide who may give voice to which of its human and nonhuman parts.

In this chapter, I begin to fill in some of the pieces that have been missing from my account of scientific authority and critique of the two-sciences settlement by attending to the role of world-building politics in creating, sustaining, and challenging the authority of political scientists and political rulers. To do so, I turn to Thomas Hobbes, whose ambition to establish a new science of politics drew him to intervene in—and to reflect deeply upon—struggles over the shape of the world and the distribution of authority within it. To a degree that many readers rightly find alarming, Hobbes sought to establish (1) a virtually incontestable sovereign authority to speak for the people and (2) his own virtually incontestable scientific authority to convince the people of the need to authorize and submit to this sovereign. When advocates of the two-sciences settlement speak of the risks of "scientizing" the study and practice of politics, they often have in mind Hobbes's twin commitments to scientific certainty and political absolutism. While such concerns are legitimate, we will see that there is more to tell about Hobbes's efforts to establish new foundations for political science than these familiar cautionary tales convey. We have much to learn from Hobbes about the conditions of possibility for both founding and disrupting the authority of political sovereigns and political scientists—lessons that may hold special value for those who lament and wish to pluralize or reshape established modes of authority in political science and political practice.

Hobbes's readers often take him at his word when he encourages us to think that he derived compulsory force for his arguments in favor of absolute sovereignty by imitating the methods of geometry, replicating the incontestable validity of a geometrical proof in political science. For example, this is Hobbes's explanation in *Leviathan* (hereafter cited parenthetically as *L*) of the stakes of making political science—and politics—more geometrical: "The doctrine of Right and Wrong is perpetually disputed, both by the Pen and the Sword: Whereas the doctrine of Lines and Figures is not so, because men care not, in that subject, what be truth, as a thing that crosses no man's ambition, profit or lust."[1] In such passages, Hobbes seems to suggest that he intends to achieve a similarly irresistible level of certainty in political science—and to prove beyond a doubt the need to consent to an irresistible sovereign—by imitating geometers' dispassionate adherence to the rules of logic. In light of this kind of argument, it is easy to see why Hobbes is portrayed as he is in much of the secondary literature: namely, as a brilliant if daft geometriphiliac who would square the circle in the morning and prove the terms of political obligation in the afternoon.[2] Critics who accuse Hobbes of "scientizing" politics fear that he is all too correct about the capacity of mathematical techniques to vanquish persuasion, discredit rhetorical appeals to the passions, and eradicate pluralistic differences of opinion about right and wrong and justice and injustice.[3] Such critics use the example of Hobbes to underscore the political dangers of scientific approaches to the study of politics and to reconfirm the two-sciences settlement's message about the need to confine mathematical methods to the study of nature, where they supposedly belong.

But such criticisms of Hobbesian science take for granted a view of the relationship between geometrical reason and political rhetoric that Hobbes's writings put into question. We will see that Hobbes, in his attempts to explain and harness the compulsory quality of geometry, does not reduce demonstration to ratiocination. As his position that geometers—and political scientists—must begin by defining their terms already suggests, Hobbes recognizes the role of language and the influence of an array of linguistic practices within geometry. Steven Shapin and Simon Schaffer are correct that, for Hobbes, "the settling of a definition was a social act," and the compulsory force of a proof depended on the success of this social act: "To end with agreement, one must start with agreement."[4] But Shapin and Schaffer only scratch the surface of the layers of agreement that support a geometrical consensus, in Hobbes's view. Hobbes himself digs deeper: as he insists in the context of his vituperative debates with other geometers of his time, to successfully settle the meaning of one word by definition presupposes the existence of other meaningful words with which to compose that definition. Linguistic meaning is not itself the product of the definitions it makes possible, according to Hobbes.

Rather, words are made meaningful—and made available for geometrical purposes and other projects of compulsion—through a quotidian, corporeal rhetoric of gestures and deeds. To give one of Hobbes's own examples, the words for numbers gradually took on meaning and acquired compulsory utility through iterative signifying practices, as people again and again performed for one another the act of touching their fingers to objects while vibrating their tongues and moving their mouths. For Hobbes, the capacity of a geometrical demonstration to compel consent ultimately depends on ongoing, embodied rhetorical praxis. A close reading of Hobbes's geometrical writings can help us to recognize that the "science effect" of geometry—its capacity to coerce consensus—is not reducible to logical validity, but rather draws crucial support from world-building rhetorical practices carried out beyond the confines of the geometer's study. When Hobbes embraces a geometrical paradigm for the study of politics, then, he does not stake the authority of his science or of the sovereign on the logical rigor of his arguments alone. Rather, read alongside his political writings, Hobbes's geometrical writings suggest that the people's creative rhetorical encounters with each other and with the other material bodies in their midst serve as a necessary condition for the emergence of coercive forms of authority in the study and practice of politics.

This point about the productive role of rhetoric in the constitution of Hobbes's new science of politics both builds on and departs from Quentin Skinner's arguments about the "marriage" between rhetoric and reason in *Leviathan*.[5] Skinner takes an important step beyond most studies of the rhetoric of *Leviathan*, which usually take for granted that Hobbes's frequent violations of his own stated prohibitions against rhetorical "abuses of words" invalidate his claims to scientific authority.[6] Skinner reminds us that, unlike today, persuasion was not generally treated as coextensive with deception in Hobbes's time. The influence of venerable humanists like Erasmus encouraged many of Hobbes's contemporaries to see rhetorical flourishes, or *ornatus*, as legitimate means to augment the political authority of a rational argument, or *ratio*. Nevertheless, Skinner still assumes that the marriage of rhetoric and reason in Hobbes's *Leviathan* is a partnership between opposites: reason provides science with stable epistemic foundations, while rhetoric creates a persuasive, public-facing façade. By contrast, I resituate Hobbes in the company of humanist figures like Lorenzo Valla, who defied the Erasmian emphasis on *ornatus* to privilege the world-constitutive impact of metaphorical, figurative language—its capacity to create not merely appearances, but the lived, embodied relationships among human and nonhuman beings that permit the reality under discussion to emerge or manifest itself.[7] Hobbes, like Vico, worked at the intersection of the maker's knowledge tradition and the rhetorical tradition, and he anticipated aspects of Vico's account of the enabling contributions of rhetorical praxis to the sciences. The act of knowing, for Hobbes, was a continuation of the act of making; in this respect, I argue, rhetorical world-building practices play a role in constructing foundations, not just façades, for Hobbes's new science.

But unlike Vico, and similarly to Descartes, Hobbes was interested in harnessing the creativity of the people to secure their exclusion from the scientific activities and authority that he would reserve for a narrow elite. Hobbes's geometrical politics requires the members of the commonwealth to be actively involved in creating the conditions under which sovereign claims about justice—and scientific claims about sovereignty—would be beyond debate, acquiring the "science effect" of a compulsory truth. Hobbes enlists the people and their world-building rhetorical practices to accomplish a feat that rational proof alone cannot: namely, to manufacture a world in which the sovereign's claims about justice and the scientists' claims about sovereignty would become as irrefutable as the conclusion that two plus two equals four. The compulsory quality of Hobbesian political science and Hobbesian politics ultimately depends on the creative activities of the people, including their acts of submission. Hobbes's theory of authorization does not just ask common men to sign on the dotted line after receiving indisputable proof that their consent is the only logical choice. Rather, he enlists them as active producers of the terms of their own domination. Like Descartes's new science of nature, I argue, Hobbes's new science of politics wavers in a constitutive ambivalence between democratic openness and absolutist closure.

My reappraisal of Hobbes's worldly geometry of politics moves in three steps: First, I discuss the complexities of Hobbes's view of rhetoric as both a means of deception and a means of creation. Next I turn to Hobbes's geometrical writings and uncover his dependence on the rhetorical activity of making meaning as an enabling condition for geometers' creative and compulsory capacities. I then use this revised understanding of Hobbes's geometry to reappraise the relationship between popular consent and the authority of scientists and sovereigns in *Leviathan*. I approach covenanting as a genre of world-building rhetoric, and show how Hobbes would enlist demotic displays of consent to secure the authority of sovereignty and of his political science.

Architectural Rhetoric

A quick glance at *Leviathan* serves to remind us why Hobbes is famous for excluding rhetoric from science. He classes metaphors among the "abuses of speech" and identifies rhetorical language as a "cause of absurd conclusions" (*L* 25–26, 34). Hobbes credits geometers with developing a method to overcome rhetoric's distortions: "Seeing then that truth consisteth in the right ordering of names in our affirmations, a man that seeketh precise truth had need to remember what every name he uses stands for. . . . And therefore in Geometry, (which is the onely Science that it hath pleased God hitherto to bestow on mankind) men begin at settling the significations of their words" (*L* 28). The implications for political science seem clear: scientists require definitions to exclude rhetoric at the outset of their reasoning in order to avoid drawing absurd conclusions at reason's end. Geometry seems to represent to Hobbes a hygienic, ascetic mode of language use and a means to guarantee the epistemological integrity of the science of politics.

But while *Leviathan* frequently draws attention to the dangers of figurative, imaginative language in science and in politics, Hobbes does not draw the strict lines between rhetoric and science or between imagination and reason that readers often impute to him. He clearly rejects the view that rational thinking begins with the triumph of reason over the imagination. For Hobbes, reason is not an immaterial faculty distinct from or menaced by the "sensuous" parts of the soul; rather, reasoning is one of the ways that the embodied imagination moves from one image to another: "I do indeed conceive that deliberation is an act of imagination or fancy; nay more, that reason and understanding also are acts of imagination, that is to say, they are imaginations."[8] Of course, Hobbes is deeply attuned to the risk that this dependence on imagination makes all reasoning prone to deception, distortion, and error. But it would be nonsensical, given his view of imagination as the material substrate of all thinking, to respond to these risks by seeking to purify reason of imagination. The imagination is the very condition of possibility of rationality, for Hobbes.

Nor does Hobbes draw strict distinctions between the imagination's rhetorical and rational deployments. Instead, he theorizes reason and rhetoric as the two "intellectual virtues" of "wit," which he defines as the speed at which the imagination is able to run through chains of past and present sense impressions (*L* 50). Hobbes's name for the rhetorical imagination is "fancy," the capacity of a quick wit to notice relationships of similarity among the images that continually move through the mind's matter. The other virtue of wit is "judgment": where a good fancy notices similarities, a good judge notices differences (*L* 51). Though Hobbes sometimes casts judgment—fancy's "severer sister"—as a check against the "Madnesse" of an unhindered fancy, he does not oppose the two, but figures them as siblings.[9] Judgment does not call to a halt the rapid comparative activity of the wit, but rather uses comparison for a different purpose than fancy. The quickest wits excel at both forms of comparison.[10] For Hobbes, the rhetorical imagination and judgment are twinned talents of agile imaginations.

Fancy's most distinctive attribute is its productivity, which helps to account for both the risks and the promise that Hobbes associates with its rhetorical products. Fancy is the more fertile of the two sisters, generating relationships among sensuous particulars plucked from disparate moments in the chain of sensations. Although Hobbes theorizes the imagination as reproductive, limited to re-creating images of things that were already experienced through the senses, fancy creates *relationships* among these reproduced images. This creativity is integral to Hobbes's understanding of the capacity of rhetorical language to persuade and deceive, but persuasion and deception do not exhaust the uses or the pleasures he finds in fancy: similitudes created by a particularly quick wit "please not only by illustration of his discourse, and adorning it with new and apt metaphors; but also, by the rarity of their invention" (*L* 51). Rhetorical invention is valued by Hobbes as a means of persuasion and as an end unto itself.

It doesn't take a great wit to notice that fancy as Hobbes defines it is similar to *ingenium* as we encountered it in Vico. "Wit" was a conventional translation of *ingenium* in Hobbes's context.[11] For Valla before Hobbes and for Vico after him, *ingenium* names the capacity to invent relations of likeness where none are given or necessary by nature. *Ingenium*, in this tradition, is both a faculty for eloquence and a faculty for apt discovery or insight, with an important role to play in both persuasion and knowledge.[12] As we saw in Chapter 2, Vico casts the creativity of *ingenium* as an underappreciated pragmatic basis for the legibility of nature and the authority of natural science; as he puts it, "from the same founts from which well-equipped speakers spring, there also might come forth the best [scientific] observers," such that even "physics can be advanced . . . through the cultivation of *ingenium*."[13] Hobbes anticipates Vico in this regard, turning to the rhetorical tradition not only for techniques of persuasion, but also for insight into the making and knowing of metaphorical relationships.

The division of labor proposed by Hobbes between fancy and judgment sometimes does consign the rhetorical imagination to an ornamental role, as Skinner leads us to expect. For instance, in the study of history, Hobbes argues, "the Judgment must be eminent," while "Fancy has no place but onely in adorning the style" (*L* 51). Hobbes advises historians to have their judgment take the lead, assigning rhetoric to the secondary role of lending an appealing style or appearances to historical judgments. But Hobbes does not always consign fancy to this decorative role. Under certain conditions, he thinks, fancy plays a productive part at an early, formative stage in ratiocination: "In demonstration, in counsel, and all rigorous search of truth, judgment does all; except sometimes the understanding have need to be opened by some apt similitude; and then there is so much use of fancy. But for Metaphors, they are in this case utterly excluded. For seeing they openly professe deceit; to admit them into Councell, or Reasoning, were manifest folly"

(*L* 52). Fancy is sometimes needed to initiate the activity of understanding. In these cases, unlike in history, Hobbes does not ask fancy to step in after the fact to adorn the finished work of the judgment. Rather, he points to circumstances when fancy is needed to *open* the understanding and begin the demonstration. In such cases, fancy's products serve as inaugural conditions for the work of judgment, not merely as supplemental adornments to make the products of judgment more palatable or persuasive. Granted, even as Hobbes assigns the rhetorical imagination to this role in the pursuit of truth, he also warns of the risk of deceit. But he would manage this risk not by excluding fancy from science but by drawing finer distinctions among fancy's figural products: similitudes are a worthwhile rhetorical risk, while metaphors cannot be hazarded because they "professe deceit." Elsewhere, Hobbes is more permissive toward "Metaphors, and Tropes of speech," arguing that they are "less dangerous" to ratiocination than other abuses of speech precisely "because they profess their inconstancy" (*L* 31). Whether metaphors or similitudes are his preferred rhetorical form, Hobbes does call for rhetorical language to be employed in demonstration, selectively and with due caution, "sometimes."

Under what conditions would the understanding "need to be opened" by fancy, requiring reasonable men and seekers of truth to depend on rhetoric despite the risks? Hobbes provides one response to this question in the *Answer to Sir William Davenant*, written shortly after *Leviathan*:

> So far forth as the fancy of man has traced the ways of true philosophy, so far forth it hath produced very marvelous effects to the benefit of mankind. . . . Whatsoever commodity men receive from the observation of the heavens, from the description of the earth, from the account of time, from walking on the seas; and whatsoever distinguisheth the civility of Europe, from the barbarity of the American savages; is the workmanship of fancy, but guided by the precepts of true philosophy. But where these precepts fail, as they have hitherto failed in the doctrine of moral virtue, there the architect Fancy must take the philosopher's part upon herself.[14]

As long as philosophy's "precepts" succeed, Hobbes would subordinate fancy to other, more reliable modes of thinking. Even as he praises fancy for devising practical, civilizing uses for philosophical truth, Hobbes still confines fancy to a subordinate role—like a workman who must follow the lead and instructions of his superiors. However, when philosophy's precepts fail, the situation changes dramatically. Fancy ceases to be an ancillary workman and becomes an architect, taking "the philosopher's part upon herself." In such cases, Hobbes adopts a chain of command akin to that advocated by Valla: "Philosophy is like a soldier or a tribune under the command of oratory, the queen."[15] As Hobbes promotes fancy to the eminent position in the search for truth, he, like Valla, indicates that the rhetorical imagination sometimes provides philosophy with strength and direction, not mere ornamentation. Indeed, by personifying the feminized fancy as an architect rather than a queen, Hobbes's adaptation of Valla's simile suggests etymologically that fancy, the creative sister, is sometimes responsible for constructing *archai*—beginnings or foundations—to support reason's projects.

What does fancy in her architectural guise build? What kind of supports do her constructions provide to demonstration? One place to turn to better understand the productive, foundational role that Hobbes envisions for the rhetorical imagination in demonstration—and compulsion—is his treatises on geometry. If geometry is the model for a political science that would be capable of securing an unwavering consensus in favor of absolute sovereignty, as *Leviathan* prominently affirms, then what is the role of rhetorical language and creativity in geometrical demonstration? Addressing this question will help us to reframe the question of rhetoric's role in *Leviathan*.

Hobbes's Geometrical Paradigm Revisited

Despite his notoriety as a geometer of politics, Hobbes is not often remembered for his contributions to the study of geometry. This is probably justified. As his rivals enjoyed pointing out, Hobbes's confidence as a geometer was out of proportion to his competence.[16] Nonetheless, oblivious to his shortcomings, Hobbes goes on the offensive against his critics in the *Six Lessons to the Egregious Professors of the Mathematics* (hereafter cited parenthetically as *SL*), where he disputes the conclusions and artfully insults the characters of the titular "egregious professors," especially Oxford mathematician and parliamentary loyalist John Wallis. For readers familiar with *Leviathan*'s portrayal of geometers as agreeable, nonpartisan pacifists, *Six Lessons* offers a surprising glimpse of the geometer Hobbes at his most combative. While this text would be poor instruction for students of geometry today, it reveals Hobbes's understanding of the role of rhetorical and political praxis in scientific compulsion in ways that are instructive for interpreters of his political theory.

In the context of his combat with Wallis, Hobbes calls on geometers to emulate political actors, not, as in Leviathan, the other way around. In one of the best-known passages from the *Six Lessons*, Hobbes likens the geometrical method of demonstration to the founding of commonwealths:

> Of arts, some are demonstrable, others indemonstrable; and demonstrable are those the construction of the subject whereof is in the power of the artist himself. . . . Geometry therefore is demonstrable, for the lines and figures from which we reason are drawn and described by ourselves; and civil philosophy is demonstrable, because we make the commonwealth ourselves. But because of natural bodies we know not the construction, but seek it from the effects, there lies no demonstration of what the causes be we seek for, but only of what they may be.[17]

Hobbes, like Vico, thought that men can only really know what they can make.[18] Hobbes argues that geometers and civil philosophers can demonstrate from causes, and thus compel belief, while natural philosophers can only conjecture about causes, and cannot demand agreement with their conclusions. The distinction he draws between the study of geometrical and political constructions, on the one hand, and natural givens, on the other, raises the question of the "power" that permits human "artists" to construct some kinds of subject matter but not others. As we have already seen, Hobbes attributes a peculiar form of creativity to fancy. Reading Hobbes's geometry in light of his account of the architectural significance of the rhetorical imagination in demonstration suggests that we should pay special attention to the role of creative language in his account of how geometers make geometrical figures and secure agreement about their properties. Such an approach will help to highlight Hobbes's concerns that geometers cannot achieve the creative or the coercive potential of their demonstrations without the ongoing support of a collective, corporeal rhetorical praxis—a gestural, meaning-making activity carried out by people who would not typically be recognized as playing any kind of role in geometrical proof. For Hobbes, as we will see, the compulsory quality of geometry relies on and is rooted in demotic world-building practices, which supply geometers with the linguistic resources they need to make and know shapes and figures. Hobbesian geometry, it turns out, depends on politics every bit as much as Hobbesian politics depends on geometry—an insight that will ultimately help to unsettle received views of Hobbes's proposal to establish more coercive forms of politics and political science.

Geometry's Rhetorical Roots

Six Lessons significantly amends the familiar view of Hobbes's geometry as a paradigm for logical purity and an exemplar of the dangers of rhetoric to science. In this text, Hobbes argues that geometers can only demonstrate truth if they first use definitions to *create* shapes and figures: "Where there is place for demonstration, if the first principles, that is to say, the definitions, contain not the generation of the subject, there can be nothing demonstrated as it ought to be" (184). To serve as a viable first principle, a definition must do more than just settle the significance of a word; it must use words to constitute the shape or figure under investigation. The first principles of Hobbesian geometry are, in this sense, "figurative": they create a figure by generating spatial relationships that did not exist prior to the act of defining them.

Hobbes even warns that undue focus on settling the meaning of words can stand in the way of geometers' creativity, undermining their demonstrations: "It is not true that every definition is a principle. For a man may so precisely determine the signification of a word as not to be mistaken, yet may his definition be such as shall never serve for proof of any theorem, nor ever enter into any demonstration, such as are some of the definitions of Euclid, and consequently can be no beginnings of demonstration, that is to say, no principles" (*SL* 200). Hobbes may have idolized Euclidean logic, as his first biographer, John Aubrey, reports, but he also critiqued and sought to surpass Euclid's deficient creativity.[19] Geometers have the opportunity and the responsibility to be not merely precise, but also generative, when authoring definitions—unlike natural philosophers, who deal with entities created by God and cannot hope to do more than settle the significance of words with their definitions.[20] Defying expectations fostered by received readings of *Leviathan*, the geometer Hobbes does not seek to displace creative, figural language with the logical language of proof. On the contrary, he insists that generative language is the prior, enabling condition for logical proof—the first step in a demonstration.

Held up to this rhetorical standard for scientific demonstration, Hobbes accuses Wallis of mistaking proof for the whole of demonstration: "Egregious logicians and geometricians, that think an induction . . . fit to be received for a geometrical demonstration! . . . Therefore though your lemma be true, and by me (Chap. xiii. Art. 5) demonstrated; yet you did not know why it is true" (*SL* 308). Wallis, Hobbes complains, "not understanding the ground of [his] profession," focuses on induction at the expense of first principles, and thus demonstrates nothing (*SL* 310). Contra Wallis, Hobbes insists that even the most rigorous logic is not self-grounding, but depends on a prior, generative step.

He further underscores the priority of generative language to logical language with an organic metaphor: "It is in sciences as in plants; growth and branching is but the generation of the root continued . . . The unsoundness of the branches are no prejudice to the roots, nor the faults of theorems to the principles. And active principles will correct false theorems if the reasoning be good; but no logic in the world is good enough to draw evidence out of false or unactive principles" (*SL* 188). According to Hobbes's simile, theorems are like branches; they grow from and depend on definitions, which are generative and supportive like roots. Proofs, like plants, live or die based on the generative activity of their root principles. If a principle fails to generate, like a dead root, logic cannot save the demonstration. Revisiting the problem of "failed precepts" in this passage, Hobbes suggests that success and failure in the search for truth entails more than mere truth or falsity: Wallis failed not because his lemma was false but because its roots were shallow and inactive. A demonstration, as Hobbes uses this term, derives its "science effect" not only from the epistemological rigor of the steps followed by its author but also from his or her creative potency.

Hobbes's surprising argument that creative, figure-making language is the root of geometrical demonstration offers a way to conceptualize geometry's first principles as products of a genre of rhetorical language. Hobbes's dispute with Wallis suggests that fancy, in its architectural guise, plays a role in the authorship of geometry's *archai*—a very different vision of the marriage between reason and rhetoric than Skinner extracts from Hobbes. Skinner never considers the possibility that Hobbes's geometry could have a rhetorical dimension, because he takes for granted that geometry, unlike political science, can compel belief through *ratio* alone. But the productive first principles that Hobbes says are necessary for a geometrical demonstration to succeed are not reducible to—and are the enabling condition for—the logical reason that follows. As Hobbes understands it, Wallis's demonstrations fail to persuade him because their roots are insufficiently active or fecund—a problem that neither better *ratio* nor more attractive *ornatus* could solve.

Hobbes's principles are rhetorical in the sense that they create the geometrical figures to be known, giving geometry its foundations, not its façade. As Hobbes's root-and-branch metaphor vividly suggests, fancy offers depth, support, and life to geometrical arguments, not mere surface appearance.

Geometry's Demotic Grounds

Digging deeper, if active rhetorical roots are required to sustain geometry's growth, from where do these roots derive their support and nourishment? If geometers' definitions exemplify the human capacity to use words as causes, not just as signs or markers, from what source do words draw this potency? Hobbes begins to address these questions by further elaborating his organic metaphor: "He that will learn geometry must understand the terms before he begin. . . . For words understood are but the seed, and no part of the harvest of philosophy. . . . [A]ll definitions proceed from common understanding; of which, if any man rightly write, he may properly call his writing *philosophia prima*, that is, the seeds, or the grounds of philosophy. And this is the method I have used" (*SL* 225–26). Even first principles have prior conditions; in order to create a figure with words, the words from which the definition is composed must first be understood. Linguistic understanding, Hobbes recognizes, cannot be created by definition: the meaning of a word is the "seed," not the "harvest" of the philosopher's activity. Hobbes cuts short the infinite regress that would result from asking definitions to give meaning to the words that make up definitions by invoking an extraphilosophical resource as the ultimate basis for philosophy. "*Philosophia prima*," the ground supporting first principles, is "common understanding," the store of shared, conventional meanings that allow a definition to be "understood in understanding the language wherein it is set down" (*SL* 229). Hobbes casts common understanding as the germ from which geometry's roots sprout or, alternatively, the soil from which the geometrical plant, root and branch, gets its support. The making of meaning, he argues, must precede the making of figures by definition.

Hobbes repeatedly objects to the idea that meaning can be constituted simply by inscribing a definition in a book. Throughout his corpus, he expresses special contempt for thinkers in any branch of learning who would abandon vernacular language and its conventional meanings in favor of obscure, made-up jargon. One of the two sources of "insignificant" language that Hobbes flags in *Leviathan* is the abundant crop of new words "coined by Schoolmen, and puzzled philosophers" (*L* 30). By contrast with these philosophers, "the common sort of men seldome speak Insignificantly, and are therefore, by those other Egregious persons counted Idiots" (*L* 59). Hobbes, of course, also has complaints about abuses of words perpetrated by common men, as we will explore in more detail shortly. But far from considering common men to be idiots, he considers their linguistic practices to be the origin of meaning and the basis for understanding. Whatever his concerns about the fallacies and instabilities of "common understanding," Hobbes utterly rejects the notion that seekers of truth should abandon common understanding in the name of correcting or stabilizing it. While he affirms the use of definitions to clarify whatever may remain ambiguous about the conventional use of words, he considers it the ultimate idiocy to reject demotic language entirely in favor of words with no vernacular meaning.[21]

To Hobbes, nothing better exemplifies this "egregious" disdain for conventional significations than algebra, an approach to geometry that would replace words entirely with what he calls "a scab of symbols" (*SL* 314). With this colorful metaphorical insult, Hobbes condemns algebra for a sin more typically associated with rhetoric: algebra's symbols are all about surfaces, he suggests, providing a misleading façade to deflect attention from underlying flaws (though where rhetorical ornaments create an appealing surface and enliven belief, algebra lends an ugly, dead appearance to a geometrical demonstration). And to Hobbes, no one better exemplifies the idiocy of preferring insignificant symbols to vernacular language than the egregious Wallis, who sets out to write demonstrations in the form of algebraic equations without even understanding what "equality" means: "[F]or liquid bodies . . . men (men, I say) measure them by putting them one after another into the same vessel, that is to say, into the same place. . . . But you (you I say), that have no definition of equals, neither received from others nor framed by yourselves; out of your shallow meditation and deep conceit of your own wits, contend against the common light of nature. So much is unheedy learning a hindrance to the knowledge of the truth, and changeth into elves those that were beginning to be men" (*SL* 221–22). Wallis's geometry is stunted because it is detached from the conventional signifying practices that make a word like "equality" meaningful and enable the growth of the geometrical plant. What Hobbes calls "the light of nature" in this passage is a *second* nature, created by men—men, he says!—when they interact with one another and with the liquids, vessels, and other bodies they encounter in the course of day-to-day life. Wallis's "deep conceit" is to rely on his own wit in isolation from these collective signifying practices.

The passage suggests the existence of creative powers that cannot be confined within the embodied mind of Hobbesian "man," because they can only emerge through the public, embodied enactments of men and women, in the plural. Where a witty thinker adeptly relates similar but disparate sensations recorded in his or her imagination, the wit exercised by these ordinary people is enacted, concerted, and public: they establish an equivalence between various different "liquid bodies" by physically pouring them into the same vessel, again and again, and then acting as though they were, in an important respect, the same: for example, pricing milk by the unit of volume. In *Leviathan*, Hobbes makes a similar point about the origins of numbers, without which "there is no possibility of reckoning." Before number words were invented, "men were fain to apply their fingers of one or both hands, to those things they desired to keep account of and thence it proceeded, that now our numeral words are but ten" (*L* 27). A shared, conventional practice of matching fingers to things, physically relating them despite their differences, constituted the meaning of quantity. Hobbes points in these passages to a primordial, deep layer of rhetorical praxis at work in geometrical reasoning: his examples help to reveal the origins of significance in the shared, repeated, mutually acknowledged gestures performed by ordinary people in their quotidian interactions. The words that geometers require in order to make and know shapes and figures by definition, like "equality," are *made* significant through this demotic, corporeal rhetorical practice.[22] Any demonstration, Hobbes suggests, derives much of its compulsory quality from the common understanding constituted through the practices of ordinary people.

Logic, on this view, is necessary but not sufficient to secure agreement to the conclusions of one's reasoning. Before the logic of a geometrical proof can even get under way, two kinds of creative acts must take place: the geometer must author appropriately active and generative first principles by definition and, before that, nongeometers must establish the meaning of the words used in these definitions. As William Sacksteder puts it, for Hobbes, "logic is not a perfection of usage. Rather usage is the matrix engendering anything we may safely call logical."[23] The meanings created and sedimented through demotic, embodied rhetorical praxis are the soil that surrounds and supports geometers' definitions and enables proofs to branch off securely from these principles. For Hobbes, the capacity of a geometrical demonstration to bind belief is ultimately dependent upon a rhetoric of deeds carried on by common men outside the confines of the geometers' study. Geometers who eschew conventional meanings uproot their proofs from the soil they need to flourish, a shortcoming again exemplified by Wallis: "You demonstrate nothing to anybody but those who understand your symbolic tongue, which is a very narrow language. If you had demonstrated it in Irish or Welsh, though I had not read it, yet I should not have blamed you, because you had written to a considerable number of mankind, which now you do not" (*SL* 261). Wallis's narrow symbol-language has no demonstrative force, for it cannot demonstrate *to* any concrete language community. For words to work as causes, they must speak to a significant portion of mankind, utilizing a vocabulary that grows from common people's practical activities. The creativity of a geometer's rhetorical imagination (and with it, the compulsory quality of his geometrical proofs) depends on the demotic rhetorical practices of a broader linguistic community—a community bound together, in turn, by the shared meanings generated by their practices. Lacking a foundation in these conventional signifying practices, Wallis's geometry is groundless, for the ultimate basis of geometrical demonstration is established by the people.

Six Lessons thus challenges us not only to rethink Hobbes's understanding of geometrical compulsion, but also his views of compulsion in politics and political science. In this text, he portrays geometers as combative and out of touch with common sense about measure and propriety, while he casts common men as agreeable and capable of giving proportion and quantity to the world. Hobbes's geometrical writings cast the activities of ordinary people as an essential resource for demanding agreement and securing peace among geometers. If this seems like a dramatic reversal of roles compared with *Leviathan*—where Hobbes laments the state of war created by conventional meanings and activities and advocates new forms of geometry-like authority to secure the peace—perhaps this is in part because we have been reading *Leviathan* with an incomplete and distorted understanding of Hobbesian geometry in mind. If geometry's compulsory qualities are, as Hobbes suggests, related to both scholarly and popular rhetorical activities, what does this mean for Hobbesian politics and political science?

The Rhetorical Grounds of Political Obligation

Of course, there are good reasons to read *Leviathan* as criticizing common men and their conventional signifying practices for failing to stabilize the meanings of words or to secure lasting agreement. *Leviathan* is full of reminders of just how *dis*agreeable ordinary people are and how difficult it is for them to overcome their diffidence. Where *Six Lessons* affirms common men and women for coalescing around a shared understanding of "equality," *Leviathan* warns of their propensity to violate agreements whenever it serves their private interests, to "grieve" one another with words, and to compete—sometimes violently—over their status as equals (*L* 86–89). *Leviathan*'s frequent complaints about the destabilizing effects of day-to-day abuses of words by ordinary people—including their casual reliance on metaphors and other tropes—complicate Hobbes's confidence in the stabilizing role of conventional linguistic practices in his discussion of geometrical demonstration in the *Six Lessons*.

Yet a closer look at *Leviathan* also complicates received views of Hobbesian political science as a way to silence the diffident multitude and force them to agree to compulsory truths—the view of *Leviathan* that underwrites many critical appraisals of "scientized" politics in debates about the two-sciences settlement. In this section, I foreground these complications in order to finesse our understanding of the absolutist authority that Hobbes's political science describes and seeks to engender. As we have seen, Hobbes's own writings on geometry place in doubt the notion that scientific demonstration requires displacing rhetoric with reason, instead casting both the rhetorical imagination and popular rhetorical praxis as enabling conditions for reasoned demonstration. On closer examination, *Leviathan*'s assertions of authority resonate with many of the lessons I extracted from the *Six Lessons*: Hobbes's new science of politics would animate—but also channel and redirect—the embodied rhetorical practices of the people. *Leviathan* offers a vision of demotic world-building politics as the generative seed and supportive soil from which an absolute sovereign—a new first principle for both politics and political science—could spring. If politics is to be scientized, *Leviathan* suggests, this will require the ongoing, active involvement of the people in constituting rhetorical, worldly foundations for their own domination.

This rhetorical dimension of Hobbes's political science does not appear clearly within the frame that Skinner and other scholars of the rhetoric of *Leviathan* have brought to the text. Based on his understanding of rhetoric as a formal technique of persuasion, Skinner centers his discussion on the tropes authored by the writer of *Leviathan*, arguing that Hobbes resorts to figurative language to make his conclusions more persuasive to audience members who are too irrational or self-interested to be compelled by logic alone. Skinner takes for granted that the need to consent to the sovereign is a *logical* necessity, discovered by Hobbes using his powers of reason and subsequently communicated using his powers of persuasion. Skinner thus does not entertain the possibility that other kinds of voices, exercising other kinds of rhetorical aptitudes, play a role in creating the demonstrable necessity of consent. On my reading, *Leviathan* does not seek to convince people to consent to circumstances that have been proven to be inevitable; rather, in this text, Hobbes casts the people's generative rhetorical activities as powerful sources of the artificial relationships that make consent obligatory. Hobbes enlists the people as agents of their own domination by an overawing authority, not mere signatories to it.

Of Hobbes's two demonstrative sciences, it is far and away more difficult to secure agreement and achieve the "science effect" in political science than in geometry. Where geometers have the luxury of beginning with words already accepted and understood and creating a figural reality from them, scientists of politics study a worldly reality characterized by continual turmoil over the meaning of certain key words: "right," "wrong," "good," "bad," "just," "unjust," "mine," and "thine" "are in the common discourses of men of inconstant signification" (*L* 31). Different men mean very different things by "justice," depending on their particular desires and aversions. For this reason, Hobbes concludes, "such names can never be true grounds of any ratiocination" (*L* 31, cf. 39, 90). Such passages echo the argument of the *Six Lessons* that words may only enter into scientific demonstrations if their significance is first created and secured through quotidian rhetorical practices. But Hobbes's point in *Leviathan* is that "justice" *cannot* be understood simply by understanding the language in which it is set down—its meaning is conventionally contested, colored differently each time it is uttered by a "tincture of our different passions" (*L* 31). When the conventional meaning of a word is caught up in a crisis of meaning, it cannot serve as a seed or ground for a demonstration, and it is as useless for authoring a compulsory argument as the philosophers' meaningless jargon or Wallis's scab of symbols. Where *Six Lessons* urged geometers to avail themselves of established conventional understandings, *Leviathan* cautions that both political order and political science are stymied by the misunderstandings created by conventional misuses of language.

Hobbes underscores that these crises of meaning cannot be solved by using definitions to clean up the messy meanings of contested terms. Scholars who have tried this approach have only deepened the problem, tincturing their definitions with their passions: "They make the rules of *good* and *bad*, by their own *liking* and *disliking*: by which means, in so great diversity of taste, there is nothing generally agreed on; but every one doth (as far as he dares) whatsoever seemeth good in his own eyes, to the subversion of commonwealth" (*L* 461). Instead of settling the meanings of contested terms, these misguided philosophers exacerbate the problem of unsettlement. Hobbes uses their failure to remind us why this problem is so intractable: words like good, bad, just, and unjust are not merely ambiguous, but meaningless, for the idiosyncrasies of the passions hijack the very signifying practices that would otherwise vest words with stable meanings. To hazard a definition for a word that *cannot* be understood simply by understanding a vernacular language is to mistake the seed of philosophy for its harvest. Scientists of politics who would bypass convention by stipulating the meaning of "justice," "good," or "mine" by definition are both egregious and superficial, in Hobbes's view.

Hobbes's assessment in *Leviathan* of the failure of popular and learned practices to constitute a stable consensus about the meaning of "justice" helps to explain his diagnosis in the *Answer* of moral philosophy's problem of failed precepts: justice, goodness, right, and other normative terms fail as first principles because their meaning has been unsettled by the ongoing signifying practices of both ordinary people and philosophers. Reading *Leviathan* in light of *Six Lessons*, it seems that, in order for arguments about justice and injustice, right and wrong, and mine and thine to become as compulsory as a geometrical demonstration, a new first principle would be required. As we have seen, these are just the circumstances in which Hobbes would give fancy a leading, architectural role in demonstration, asking the rhetorical imagination to open the understanding and invent new *archai*.

Even as Hobbes encourages his readers to accept the failure of conventional signifying practices to create meaning for justice, he points toward a genre of rhetorical performance that promises to stabilize meaning and create the conditions of possibility for a newly compulsory consensus in politics and in political science: namely, the pronouncement of covenants. Hobbes first draws our attention to the creativity of covenanting in *Leviathan*'s introductory body-politic metaphor, in which he likens those who pronounce covenants to the author of nature: "The pacts and covenants by which the parts of this body politic were at first made, set together, and united resemble that fiat, or the let us make man, pronounced by God in the creation" (*L* 9–10). Hobbes likens the act of pronouncing covenants to the divine creation of nature; indeed, he suggests that covenanters' acts of speech, by creating unity, abolish the war that characterizes the state of nature as it was made by God. Hobbes adds another link to this chain of likenesses in the *Elements of Law*, suggesting that the God-like creative power exercised by covenanters is also like the human intellectual power responsible for creating similitudes: "Erecting a body politic . . . is like a creation out of nothing by human wit" (107). Those who covenant approximate God's ex nihilo creative power by making something similar to a similitude, a pact that sets together what is, by nature, disparate and diffident: namely, their bodies.[24] The Hobbesian social contract, on this reading, is an enacted, collective mode of rhetorical performance; it creates a newly unified constellation of political relationships among its performers, as if out of nothing. By reading Hobbes's two similitudes alongside the *Six Lessons*, we are better able to see a third resemblance: like geometers, whose definitions create shapes and figures and make possible the coercive quality of subsequent demonstrations, the covenanting multitude creates anew the reality studied by political science, paving the way for its demonstrations to be as coercive as geometrical proofs.

As Skinner emphasizes, Hobbes's repeated recourse to similitudes in his self-described science of politics belies the facile opposition between reasoned argument and rhetorical presentation that is often imputed to him. But Skinner's focus on the rhetorical form of Hobbes's composition tends to obscure the popular rhetorical powers and activities that Hobbes seeks to evoke and conjure with his compositional choices—the rhetoric of covenanting that the people, not the political scientist, put into motion. Even as Skinner seeks to move us beyond readings that oppose reason to rhetoric, he reinforces the underlying assumption that the compulsory quality of *Leviathan* comes from the words that *Hobbes* pronounces as its author. His reading of Hobbes's body politic metaphor is symptomatic of this limitation: in Skinner's view, Hobbes depicts the sovereign as the soul of the body politic in order to frighten people into submission, painting "a picture of vastness and potential violence . . . to be approached not merely with respect but with awe and reverence."[25] But this is to elide the curious doubling of divinity in *Leviathan*'s introductory metaphor: Hobbes suggests that people are most god-like as creators when their covenants create a god-like power to awe and rule them. Based his understanding of rhetoric as a means of persuasion,

Skinner takes Hobbes's metaphor as a means to induce the submission to sovereigns that logic demands, but which irrational people are prone to misunderstand. But Hobbes not only uses his own language to seduce men to consent—whether from fear, logic, or of some combination of the two. He calls upon men to use their *own* bodies and words to create this fearful figure, as if out of nothing—an embodied, world-building rhetorical praxis that is difficult to recognize as such if we conflate rhetoric with techniques of persuasion.

In contrast with the creative power that he attributes to covenanters, Hobbes dampens expectations that his own speech, as the author of *Leviathan*, could create the political artifice he describes—a marked departure from his arrogant claim to have single-handedly come to Euclid's rescue by supplying geometry with suitably generative definitions in the *Six Lessons*. *Leviathan's* "Definition of a Commonwealth" expresses deference to the multitude, whose acts bring the sovereign commonwealth into being: "The essence of the Commonwealth . . . (to define it,) is *One Person, of whose Acts a great Multitude, by mutuall Covenants one with another, have made themselves every one the Author, to the end he may use the strength and means of them all, as he shall think expedient, for their Peace and Common Defence*" (121). By definition, the commonwealth is made by covenant, not by definition. In another similitude, Hobbes tells us that this act of covenanting is "as if every man should say to every man, *I authorize . . .*" (*L* 120). This may make it sound as though the people who covenant are passively submitting to the inevitable, halfheartedly reciting a script to hand power over to the sovereign. Phillip Pettit, encouraging this view, argues that the *sovereign's* pronouncements, not the speech acts of covenanters, are the genuine source of the commonwealth's unity: "Where does the unity of the personated multitude come from, given the plurality of the materials available? From the individual who does the personating. . . . The idea is that the many, having consented to common personation, are unified in the personator, embodied in what he or she says or does in their common name."[26] But while Pettit is right to emphasize that the sovereign is the figure authorized to speak by the Hobbesian social contract, Hobbes makes clear that this authority is produced by the words and gestures of those who do the consenting. The sovereign's very capacity to represent and compel the people with his words is generated by their ongoing performances of acquiescence.

This feature of Hobbes's account is easy to overlook, given the more obviously theatrical role that he assigns to the sovereign. Like a stage actor,

Hobbes argues, the sovereign's speech acts are considered by spectators to be performances of a script authored by another (*L* 112). But, outside of the theater, it is difficult to discern when a person speaks only for himself and when he speaks on behalf of others. For this reason, Hobbes argues, those who authorize a person to represent them must make a show of their consent, as "counter-assurance" of the person's authority (*L* 113). If authors fail to perform this counter-assurance, then the person's "authority is feigned"—his words and actions may have the superficial appearance of authority, but they cannot authoritatively oblige. Like Wallis's "scab of symbols," a ruler whose authority has no grounds in the signifying practices of the people is but a feigned sovereign and an insignificant symbolic authority. Much as *Six Lessons* deferred to the common people's conventional signifying practices as the germ from which root principles grow, Hobbes's theory of personation in *Leviathan* casts the multitude's practices of signifying consent as the germ that gives rise to the sovereign's authority to compel them to do whatever he says is just.[27]

Nor is this foundational rhetoric of covenanting a one-time performance, after which the sovereign's words would be all that matter. As Samantha Frost shows, *Leviathan* is consumed with illuminating the repertory of words and gestures through which men continually show consent—what she calls the "daily covenant"—including "*Civill Worship*," "the Worship we exhibite to those we esteem to be but men, as to Kings, and men in Authority" (*L* 447).[28] For members of the multitude wondering how to create a perceptible "counter-assurance" of the sovereign person's authority, Hobbes has suggestions: "To Pray to, to Swear by, to bee Diligent, and Officious in Serving: in summe, all words and actions that betoken Fear to Offend, or Desire to Please, is *Worship*, whether those words and actions be sincere, or feigned: and because they appear as signes of Honoring, are ordinarily also called *Honor*" (*L* 446–47). Other signifying practices that betoken honor include "to give great gifts," "to flatter," "to shew any signe of love, or feare," "to praise, magnifie, or call happy," "to appear before him with decency, and humility," "to believe, to trust, to rely on," "to hearken to a mans counsel," and "to agree with in opinion" (*L* 64–65). Hobbesian covenanters do not exit the stage after reciting a brief script about authorization. Hobbes calls upon them to perform their daily lives "as if" their every breath, gesture, and utterance pronounced the words "I authorize." To generate a sovereign, the people must breathe from the belly and bend at the knees, palpably demonstrating his authority in ritualized, continual shows of submission. Like a plant from the ground, the compulsory quality of the sovereign's speech acts grows from the people's visible, corporeal demonstrations of their consent.

On this reading, covenanting is accomplished by channeling the rhetorical faculties, practices, and powers through which common people create meaning for words toward the end of creating meaningful authority—a sovereign to continually overawe the very people who continually author the conditions for his capacity to coerce. Covenanting is irreducible to submitting, relinquishing, or surrendering; as Hobbes puts it, the generation of the sovereign commonwealth is "more than consent." Covenanting is a ritualized, ostentatious world-building practice, through which disparate, diffident multitudes reshape themselves into a "real unity" and create real authority for their sovereign. The Hobbesian covenant describes a form of domination conditioned by the people's active, energetic rhetorical praxis, not their quiescence. What are the implications of this reading of covenanting as a genre of generative rhetoric for our understanding of Hobbes's political vision and broader debates about rhetoric and science in political theory?

Conclusion

First, this reading clarifies the distinctive version of absolutism advocated by Hobbes without qualifying his status as an absolutist. In the commonwealth Hobbes describes and affirms, the people's rhetorical praxis is constitutive of a form of authority that rules out their involvement in debating justice, property, and morality. Unlike the despot theorized by Montesquieu, for example, whose power depends on the somnambulance or even paralysis of his subjects, the sovereign theorized by Hobbes depends for his power on an energized, loquacious populace. Hobbesian subjects are asked not to passively resign themselves to whatever conclusions logic forces them to accept, but rather to actively channel their signifying activities in the service of creating the irrefutable reality of sovereignty. Thus, while Hobbes is a theorist of the radical creativity of demotic rhetorical practices, he is not, as James Martel has recently argued, a radical democrat.[29] On the contrary, Hobbes reminds us that popular activity can animate both democracies and despotisms, for the people are capable of making many different kinds of meaning. This point tends to get lost when we take Hobbes's mode of rhetorical address to the people as a sign that he would have them overthrow their sovereign, as Martel holds, or elide the contingency of the relationship between demotic activity and democracy, as even Frost tends to do. Hobbes can best serve as a resource for those who wish to affirm the power of the people and contest the rule of overawing sovereigns if we let him stand as a reminder of the ambivalence of popular meaning-making practices—their foundational role in both democratic self-rule and its suspension.

Second, my reading of the rhetorical, popular foundations of Hobbes's absolutism complicates the cautionary story about the risks of "scientized politics" that is often derived from his writing, and which still circulates widely in discussions about science in political theory today. In the two-sciences literature, Hobbes is taken to represent the dangerous hubris of those who would replicate the certainty of demonstrated truth in the political realm or model political rule on scientific method. In Sheldon Wolin's memorable phrase, Hobbes gives us a sovereign "Great Definer," who silences debates about justice by dictating its meaning.[30] This line of critique is not wrong—indeed, it captures a sense of Hobbes's antipathy toward democratic judgment that is missing from portrayals of the democratic Hobbes—but it is incomplete. Even as Hobbes calls for a sovereign Definer "to set forth and make known the common measure by which every man is to know what is his, and what another's; what is good and what bad; and what he ought to do, and what not," he calls on the people to set forth and make visible the authority of the sovereign. In the absence of this continual infusion of rhetorically performed assurance, the sovereign would be just as impotent to dictate the meaning of justice as ordinary people or "puzzled philosophers."[31] The Hobbesian sovereign is more like a geometer than Wolin realized: the sovereign's geometry-like capacity to compel agreement with his definition of justice ultimately springs from the meaning-making practices of the people, as plants spring from seeds or proofs from *philosophia prima*. Only on this popular basis, Hobbes suggests, can the sovereign's judgments take on the compulsory quality of a demonstrated truth. Hobbes scientizes politics by enlisting the people to accomplish a feat that proof, no matter how rigorous, cannot: namely, to manufacture the conditions under which claims about justice can be placed beyond contestation, acquiring the science effect of compulsory truths.

But third, in exposing the demotic, rhetorical underpinnings of Hobbes's antidemocratic approach to politics and science, my readings uncover resources that could be marshaled to oppose Hobbesian absolutisms. Justice may be more like equality than Hobbes realized, and the creative powers of common men and women, acknowledged by Hobbes to rival those of God, may also rival those of the sovereign. Reading Hobbes against Hobbes, we may conclude with cautious optimism that a rhetorical micropolitics of enacted meanings could offer a powerful, if uncertain, means to pursue a justice of our own making, and to remake overawing sovereigns into feigned authorities and empty symbols.

Epilogue: Science and Politics at the End of the World

This book has elaborated and defended the view that a politics of world-building is responsible for constituting the authority of those who speak for nature, underwriting scientists' enduring and controversial role as spokespersons for the voiceless, nonhuman phenomena of the material environment. An array of sensuous, corporeal, and rhetorical world-building practices suture together a web of relationships among diverse human and nonhuman bodies, traversing, defying, and defining their differences. In the course of assembling these relationships, some things are accorded and other things denied a place in the perceptible, speakable world, and some actors are granted and others denied the authority to speak on behalf of various parts of this world. Such events make up a politics of science that involves, but is not reducible to, the experiences and activities of scientists themselves. From the perspective I have advanced, the sciences are political because they are among the most important sites where moderns have struggled with and against each other to inherit, dismantle, rebuild, and preserve the world. These struggles decide when and where it is possible to raise or answer—sometimes with hope, sometimes desperation—the question posed in my title, *Who Speaks for Nature?*

Today, faced with the twin dangers of climate change and climate denial, it has never been more important to understand the conditions under which the authority to speak for material phenomena is produced, eroded, and reinvented. The fossil fuel industry, seeking to protect its profits at the expense of the earth, has invested hundreds of millions of dollars to cast doubt on what scientists tell us about the human causes and catastrophic effects of climate change. Such campaigns to deny the reality of climate change perversely highlight the unique authority of science by seeking to destroy it. The words of scientists matter enough for special interests to care to suppress them because they are, as we have seen, intimately connected with the passions and actions of citizens and sovereigns—connections that are intricate, layered, and multidirectional. The phenomena for which scientists speak are rendered perceptible, speakable, and intelligible through multigenerational popular struggles to establish a common world from the disparate materials encountered on earth's surface. The authority of scientists depends on the power of common sense engendered in the course of these struggles, though it also depends on those who wield this power willingly channeling and selectively withholding it to constitute themselves into an eager audience for scientists' speech. The politics of science, thus conceived, creates a tissue of affinities, habits, rhetorics, and affects that hold disparate human and nonhuman bodies together in a common world and enable some of them to speak for others. Climate denial attacks not only the credibility of science, but also the integrity—physical and moral—of this tissue; it endangers the earth by endangering the world, for it is only as part of the world that the earth and its climatological changes become objects for human speech and action.

It may be tempting to think that these attacks could be defeated and that scientists' status as spokespersons for the climate and its crises could be secured on purely epistemological grounds. For example, Neil deGrasse Tyson—himself one of the most revered scientist-spokespersons of our time—answered in an epistemological idiom when asked (in an interview with Stephen Colbert) whether journalists should listen to what climate deniers have to say in the name of balanced reporting: "Once science has been established, once a scientific truth emerges from a consensus of experiments and observations, it is the way of the world. . . . That's the good thing about science: It's true whether or not you believe in it."[1] On this account, scientists establish their exclusive claim to speak for the climate by establishing, in the lab or the field, a solid evidentiary basis for the things they say in public. The good thing about science, Tyson suggests, is that, *unlike* a political claim, it need not appeal to the beliefs of a wider public or be exposed to debate to become "the way of the world." Such an analysis suggests that the evidence, not the people, decides the question of who speaks for nature. This epistemological understanding of scientific authority is, of course, not unique to Tyson; in fact, his line affirming science for eschewing popular belief has become an internet meme and T-shirt slogan because it captures sentiments about science that are widespread. In the view of many scientists and nonscientists alike, the value of science in an age of climate change consists in its practitioners' disregard for the beliefs of its admirers and its detractors.

Elevating the voices of climate scientists who speak up about the evidence they have gathered of anthropogenic climate change is indeed an indispensable element of the struggle to end our dangerous dependence on fossil fuels and establish a more just and sustainable world. But if the scientific consensus about the reality of climate change is to become the "way of the world," in Tyson's apposite phrase, it is not enough for scientists to have the truth on their side. The reality of climate change must be not merely believed but lived, breathed, and acutely felt by the people in their everyday and extraordinary acts of perceiving, desiring, dismantling, assembling, and otherwise creating a world. We must make science more, not less, political—in this specific, worldly sense—in order to respond to the most consequential ecological and political problems of our time.

Only a collective politics of world-building can rescue climate change from the fog of unreality to which all phenomena in a world of appearances are vulnerable. As Arendt helped me to argue, *common* sense, our power to feel the "realness" of appearing things, is won by establishing a pluralistic space of appearances, where spectators may witness each other in the act of perceiving the things that appear between them. From an Arendtian perspective, the best way to respond to doubts about the existence of climate change is to make it an object of public interest and controversy. As she puts it, "The world *comes into being only if* there are perspectives; it exists as the order of worldly things only if it is viewed, now this way, now that, at a given time."[2] Common sense, by grasping the undeniable sensuous presence or "realness" of things that appear from multiple vantages, makes them eligible to be re-presented in speech, imagination, and judgment as part of the common world. From an Arendtian perspective, to publicize *only* the consensus views of climate scientists risks atrophying the realness of the phenomena scientists study, undermining their power to represent this reality. The problem with the journalistic convention of presenting "both" sides of the climate debate is that two sides is too few, not too many, to make the reality of the changing climate into the way of the world.

But climate change also strains the power of common sense to guarantee the world's realness in ways that Arendt herself did not anticipate, but can help to illuminate. The entropic phenomena of climate change make their appearances precisely by disrupting previously established patterns and edifices on earth. What would it take for a man-made wave of disruptions in earth's cycles of birth, death, growth, and decay to itself become part of "the order of worldly things"? As Arendt tells us, common sense can only guarantee the "sheer thereness" of appearing phenomena if they either remain still long enough to become an object of collective attention or at least derive artificial stability from a surrounding infrastructure of durable objects. But in an age of anthropogenic climate change, *nothing* remains the same as it was in the past, and we cannot expect anything we build to last into the future—a shift that has excited the apocalyptic imaginations of media producers and consumers as never before. We can no longer count on buildings, levees, dams, bridges, and other human works to withstand the elements and anchor the durational frame within which the world's realness is secured.

Nor can we rely on the earth's own seasons as a source of regularity and realness. The unintended consequence of decades of burning fossil fuels to accelerate human work and labor has been to destabilize everything on earth and destroy the last vestiges of Arendt's distinction between the given and the made. Climate change is a crisis in the very fabric of reality—a version of "earth alienation" in which nothing appears the same for long enough to become a stable node in the network of relationships that holds together the world. An Arendtian analysis of realness can help us to recognize that climate-denial propaganda is effective not only because it is well funded, but also because it echoes and exacerbates an underlying sense of unreality caused by the very climatic events it denies. The credibility accorded by large swaths of the American public to "merchants of doubt"—scientists who supply the carbon industry with specious evidence denying the human causes or harmful effects of climate change—reflects the fact that their pseudoscientific equivocations rhyme with the widespread experience of alienation from earth's increasingly fleeting, disordered appearances.[3]

In the face of this climatic crisis of common sense, the strong consensus among many, many scientists that climate change is real should itself be understood and affirmed as an accomplishment of world-building politics. We could repurpose the narrative arc of Vico's story of the origins of the "world of nations" to highlight the political grounds of this scientific consensus and complicate its casting as a purely epistemological accomplishment: Like the fictive antediluvian humans of Vico's origin story, the scientists who originated this consensus were, at first, isolated from one another, engrossed—in this case—in the different materials and methods of their separate fields of specialization. But this condition of disciplinary isolation was interrupted when unprecedented climatic events began to pull in tandem upon the needles of the various instruments scientists had trained on their various objects of study. Fluctuations in the global climate elicited newly shared feelings of fear and fascination and inspired scientists to reorient their actions and align their passions around a common object of concern. Scientists gathered together from their distant institutional outposts to constitute primordial political collectives such as the Intergovernmental Panel on Climate Change (IPCC), a union of thousands of members and reviewers hailing from dozens of different disciplines and nations.[4] Contravening Arendt's characterization of natural science as the narrowing of perspective, the IPCC inaugurated an Arendtian space of appearances. Many eyes, peering through many different lenses, perceived from different perspectives the danger looming over all manner of earthly phenomena: icebergs, disease vectors, biogeochemical balances, sea levels, crop yields, heat sinks, and so on. In a marvelous example of Vichian *ingenium*—the faculty through which good speakers and good observers create likenesses among things that are disparate or different—these scientists created their consensus by establishing unprecedented connections among diverse manifestations of earth's terrible changes. In the parlance of Vico's maker's knowledge doctrine, they participated in remaking the world to be known by giving meaning, measure, and relation to the cacophony of semblances assailing their instrument-aided senses.

The scientific consensus on the reality of climate change, conceived in these Vichian terms, is more than just an agreement about which conclusions are and are not merited on the basis of the evidence. This consensus is a web of relationships, woven by and among scientists, in the plural, and drawing together the various material phenomena they have encountered in the course of their research practices. As the etymology of consensus suggests—from the Latin *com*, with, and *sensus*, perception, feeling, undertaking, or meaning—this unity of scientists is an instance of common sense, the feeling that the things I perceive are likewise sensuously present to others, even if they see and respond to them differently. The forging of this consensus entailed the Herculean feat of holding change itself in place so that it could be regarded and responded to by more than one spectator at once as an enduring part of the common world. When the scientists who authored this consensus speak for the things that appear between them, their authority to do so derives in large part from their involvement in gathering their bodies, instruments, and data together to create a novel worldly formation where manifestations of climate change can appear.

But so, too, do the limitations of this authority derive from the obstacles that prevent most people from joining the narrow circumference of the specialists and participating in the world-building activities that give rise to their consensus—obstacles that climate change itself has reinforced. For those of us who do not regularly see the world through the sophisticated measuring instruments of the sciences, climate change is barely perceptible on a day-to-day basis, except perhaps as a nagging feeling that someone has rearranged the earth's furniture an inch to the left as a cruel practical joke. Instead of gathering us together, this uncanny feeling of earth's instability alienates us from the worldly in-between and deepens our isolation from one another. The disorientations of life on a tilt-a-whirl planet have taught us to love our isolation and avoid scenes of collective action and judgment, much like the men of "good sense" painted so vividly by Descartes in the first lines of the *Discourse on Method*. Like them, we have lost our taste for consensus and developed a preference for judging the world alone, from the narrow vantage afforded by our individual paths through it, and without concern for the perspectives of others.

Indeed, even those who wear Tyson's words on their chests exemplify, in their very affirmation of the value of science, the isolation of good sense: their slogan proclaims that they do not care what beliefs other people hold about the sciences that they revere. Their affirmative feelings for science may be widespread, but they are not yet *shared*—they are not yet a common sense, which arises from the regard that each individual has for the other people regarding the same things from different vantage points. As Descartes understood, under such conditions, scientists cannot attain authority simply by airing the evidence and expecting truth to speak for itself. The people who would be united by a shared desire to hear what climate scientists have to say do not yet exist. If the scientific consensus on climate change is to become the way of the world, this people must be founded.

It is by no means easy to see how this feat of world-building could be accomplished, but Descartes's own struggles to surmount the paradox of founding and to unite the people around a new desire for his new science are instructive. Descartes understood that, when there is nothing left in the world to attract the attention and mutual concern of more than one spectator at a time, scientists themselves have a responsibility to step forward and become that spectacle. Descartes stepped into the public eye by narrating the story of his lifelong pursuit of the truth, exciting the imagination and inviting the judgment of an inexpert reading audience. His acts of self-exposure sought to constitute a space of appearances where a people could be born, bound together by their shared desire to know what he knew, their shared fear of illness and death, and their shared faith in his promise to devote his research solely to saving lives. Today, when both life and life in common are imperiled by climate change, there is both a special need and a special opportunity for scientists to emulate Descartes and "world" themselves, stepping forward to appear as objects of shared attention, interest, and enjoyment. Tyson himself embodies this aspect of the Cartesian legacy as host of the hit science documentary series *Cosmos: A Spacetime Odyssey* and as a Twitter impresario with more than five million followers—actions that belie his own stated disregard for the opinions people form of science.

These Cartesian modes of eliciting passionate attachments to the sciences stop short of leveling the difference between experts and their eager lay audiences. Rather, this way of engendering authority for science cultivates shared desires among the people to be subjects, not agents, of that authority. Descartes, we have seen, solicited the people to enforce their own exclusion from his new science, creating the worldly conditions for him to complete his research alone and undisturbed. A Cartesian approach to the politics of climate science would call on the people to act as a firewall to protect researchers—and their funding—against interference, whether by the people themselves or by those who speak for them in government offices. Such a "partition of the perceptible," in Rancière's theoretical language, distributes the authority to speak for nature asymmetrically away from the people, but also depends on them and their common sense to effect that distribution. Even when the people mark themselves as lacking any qualification to judge the objects of nature, they may nonetheless audaciously claim the role of judging who *is* qualified and on what basis, choosing from among those who claim the part of scientist whom they would give—and whom they would deny—the public ear and the public dollar. The asymmetrical distribution of the authority to speak for nature is a world-building project that depends as much on the comportments, affects, and commitments of those who are precluded from acting as scientists as it does on scientists themselves or the evidence they discover, compile, and present.

But it is also for the people to judge whether and when such asymmetries themselves endanger life or erode the conditions for life in common and to decide whether and how to challenge, dismantle, and rework the authority to speak for nature. For example, consider the response of the people of Flint, Michigan, to their city's ongoing water contamination crisis, which began when an unelected city manager decided in 2014 to source water from the polluted Flint River as a cost-cutting measure. Flint residents could see, taste, and smell that the water was foul, though the gravest danger—lead leached from aging service pipes by the river's corrosive water—was invisible and tasteless. They organized to raise concerns and objections, which were summarily dismissed by the expert bureaucrats of the Michigan Department of Environmental Quality (MDEQ), the city manager, and Rick Snyder, the governor who appointed the manager. The water, these scientists and sovereigns falsely insisted, had been rigorously tested and was safe.[5] Flint residents responded not by deferring to MDEQ scientists but by challenging their authority to speak for the contents of the water, enlisting outside experts from Virginia Tech to contest the results of the state-sanctioned tests.

The research conducted through Virginia Tech was groundbreaking not only for its results—the tap water of more than a quarter of Flint households indeed violated the federal lead limit of fifteen parts per billion—but for its means of data collection: Virginia Tech experts collaborated directly with local water activists, training them to collect water samples from their neighbors in Flint to be analyzed in Virginia.[6] Virginia Tech scientists and Flint citizens acted together to reorganize the worldly conditions governing who may—and may not—speak for the quality of the air, water, and soil. The people of Flint did not merely call upon one group of experts to debunk and supplant the expertise of the state; they enacted a new version of expertise as a collaborative venture of the people, united by their tangible connections to the shared dangers of their material environment and to the scientists of their choosing.

The practices through which the people engage with their physical milieu underwrite—and may undermine—the authority of both scientists and sovereigns, a point that Hobbes understood full well. In the case of Flint, Governor Snyder declared a citywide state of emergency and issued an official apology for his administration's actions in the aftermath of the Virginia Tech study and the citizen organizing that brought it into being. Though he remains in office (as of this writing), he faces mounting pressure to resign. Flint residents' alliances with each other and with the Virginia Tech scientists placed the grounds of his authority in doubt. Testing the water for themselves was a way of testing his authority; the results of those tests revealed not only that the water was unsafe to drink, but that Snyder's authority to rule the city through unelected management was, as Hobbes would say, "feigned." Flint serves as both a cautionary tale of the dangers posed when scientists and sovereigns are disconnected from the communities they serve and an inspiring example of how citizens and scientists can devise new, more just and sustainable ways of aligning the activities that transpire in the laboratories and in the streets.

Like the Flint water crisis, the global climate crisis calls upon scientists, citizens, and rulers alike to experiment with new constellations of expertise, activism, and authority. Only by way of our collective actions and passions can the scientific consensus about the reality of climate change become the way of the world. In the words of civil rights pioneer Bayard Rustin, "The proof that one truly believes is in action." The world in which climate scientists have more political clout than the carbon industry has yet to be built; its creation will require that scientists, citizens, and sovereigns act together to revitalize a worldly politics of science.

NOTES

Introduction

1. Steven Epstein, *Impure Science: AIDS, Activism, and the Politics of Knowledge* (Berkeley: University of California Press, 1998), 1. In the case of Woburn, citizens' efforts to document clusters of illness around toxic waste sites were dismissed by the Massachusetts Department of Public Health, the Centers for Disease Control, the Environmental Protection Agency, and the American Cancer Society "as the work of amateurs"; "The harshest criticisms were directed at the very idea of public participation in science" (Frank Fischer, *Citizens, Experts and the Environment: The Politics of Local Knowledge* [Durham, NC: Duke University Press, 2000] 153, 155).

2. For example, 350.org affirms that "science is on the agenda" of their organization and calls on activists to make sure the president (then Obama) "not only listens to the science, but really hears it." https://350.org/science-agenda, accessed January 19, 2017.

3. Annual funding to climate-denial organizations amounted to $900 million between 2003 and 2010. The sources of this money are increasingly untraceable, passing through charitable foundations, which are not required to report their donors publicly. For example, although ExxonMobil was "heavily involved" in climate denial earlier in the decade, their last publicly transparent donation to the cause was in 2008. Robert J. Brulle, "Institutionalizing Delay: Foundation Funding and the Creation of U.S. Climate Change Counter-Movement Organizations," *Climatic Change* 122 (2014): 681–94.

4. My use of "the science question" here and in the title of this Introduction pays homage to Sandra Harding's landmark *The Science Question in Feminism* (Ithaca, NY: Cornell University Press, 1986).

5. The epistemological framing of deference to climate science can be seen, for example, in the conversion to an internet meme of a recent quote from the esteemed astrophysicist Neil deGrasse Tyson: "The good thing about science is that it's true whether or not you believe in it." I return to this example in the Epilogue. On the other side, we can see a similar conflation of science's authority with its degree of validity in the testimony of climate denier Scott Pruitt at his confirmation hearing to helm President Trump's Environmental Protection Agency: "I believe the ability to measure with precision the degree of human activity's impact on the climate is subject to more debate on whether the climate is changing or whether human activity is contributing to it." Andrew Rafferty, "Scott Pruitt, Trump's Pick to Head EPA, Questions Human Impact on Climate Change," *NBC News*, http://www.nbcnews.com/politics/politics

-news/scott-pruitt-trump-s-pick-head-epa-faces-questions-climate-n708361, accessed January 19, 2017.

6. I have in mind such groundbreaking challenges to positivist epistemology as Paul Feyerabend, *Against Method* (New York: Verso, 2010); Thomas Kuhn, *The Structure of Scientific Revolutions* (Chicago: University of Chicago Press, 1996); and Imre Lakatos, "Falsification and the Methodology of Scientific Research Programmes," in *Criticism and the Growth of Knowledge*, ed. Imre Lakatos and Alan Musgrave (Cambridge: Cambridge University Press, 1970).

7. To name but a few prominent examples from this robust and diverse field: Karin D. Knorr-Cetina, *The Manufacture of Knowledge: An Essay on the Constructivist and Contextual Nature of Science* (Oxford: Pergamon Press, 1981); Anne Fausto-Sterling, *Sexing the Body: Gender Politics and the Construction of Sexuality* (New York: Basic Books, 2000); Sandra Harding, ed., *The "Racial" Economy of Science: Toward a Democratic Future* (Bloomington: Indiana University Press, 1993); Londa Schiebinger, *The Mind Has No Sex? Women in the Origins of Modern Science* (Cambridge, MA: Harvard University Press, 1991); Londa Schiebinger, *Plants and Empire: Colonial Bioprospecting in the Atlantic World* (Cambridge, MA: Harvard University Press, 2004); and Steven Shapin, *A Social History of Truth: Civility and Science in Seventeenth-Century England* (Chicago: University of Chicago Press, 1994).

8. Lorraine Daston, "Science Studies and the History of Science," *Critical Inquiry* 35 (2009): 801.

9. I am grateful to Samantha Frost for pressing me to think through this seeming incongruity.

10. Sophia Rosenfeld's *Common Sense: A Political History* (Cambridge, MA: Harvard University Press, 2011) elaborates the persistence of the view of common sense as convention in the history of modern political ideology while also documenting consequential moments when common sense's meaning was creatively repurposed as a quotidian faculty for insight into basic realities and truths.

11. Hannah Arendt, *The Life of the Mind* (San Diego: Harcourt, 1978), 51.

12. Perhaps the only sustained engagement with STS in political theory to date is Mark Brown's illuminating *Science in Democracy: Expertise, Institutions, and Representation* (Cambridge, MA: MIT Press, 2009). I return to Brown's contribution to STS and political theory in the second section below. Lisa Disch also helps to initiate a dialogue between STS and political theory in "Faitiche'-izing the People: What Can Radical Democratic Theory Learn About Political Representation from Science Studies," in *The Stuff of Politics*, ed. Bruce Braun and Sara Whatmore (Minneapolis: University of Minnesota Press, 2010).

13. Karen Barad, *Meeting the Universe Halfway: Quantum Physics and the Entanglement of Matter and Meaning* (Durham, NC: Duke University Press, 2007), 132.

14. Gayatri Chakravorty Spivak, "Can the Subaltern Speak?," in *Marxism and the Interpretation of Culture*, ed. Cary Nelson and Lawrence Grossberg (Urbana: University of Illinois Press, 1988), 308.

15. Mark Reinhardt, "Who Speaks for Margaret Garner? Slavery, Silence, and the Politics of Ventriloquism," *Critical Inquiry* 29 (2002): 84.

16. Ibid.

17. The category of nature, thus defined, is not "natural," if by this we mean essential or given prior to human intervention, but a "second nature" (as Crina Archer, Lida Maxwell, and I have called it) produced, maintained, and contested through the boundary work of building the world. Archer, Ephraim, and Maxwell, "Introduction: Politics on the Terrain of Second Nature," in *Second Nature: Rethinking the Natural Through Politics* (New York: Fordham University Press, 2013).

18. *Project Nim*, directed by James Marsh (2011; United Kingdom: BBC Films, 2012), DVD.

19. As Chapter 3 makes clear, my use of the phrase "constituent moments" is informed by Jason Frank's illuminating study of the dilemmas of authorization that arise when an appeal to "the people" for authorization must be vindicated retrospectively. *Constituent Moments: Enacting the People in Postrevolutionary America* (Durham, NC: Duke University Press, 2010).

20. Kimberley Curtis, *Our Sense of the Real: Aesthetic Experience and Arendtian Politics* (Ithaca, NY: Cornell University Press, 1999); Patchen Markell, "Arendt's Work: On the Architecture of *The Human Condition*," *College Literature* 38 (2011): 15–44; Ella Myers, *Worldly Ethics: Democratic Politics and Care for the World*, (Durham, NC: Duke University Press, 2013); Kelly Oliver, *Earth and World: Philosophy After the Apollo Missions* (New York: Columbia University Press, 2015); and Lena Zuckerwise, *The World Is Plural: Hannah Arendt and Democratic Politics* (forthcoming).

21. Mr. Perestroika's email is reprinted in Kristen Renwick Monroe, ed., *Perestroika! The Raucous Rebellion in Political Science* (New Haven, CT: Yale University Press, 2005), 9–11.

22. Sheldon Wolin, "Political Theory as a Vocation," *American Political Science Review* 63 (1969): 1064.

23. Similarly, Hannah Arendt charged proponents of quantification in political science with "the willful obliteration of their very subject matter," for the contingent deed, which she considered the essence of politics, is "ruled out as immaterial" by sciences that treat knowledge as coextensive with the measurement of predictable trends (*The Human Condition* [Chicago: University of Chicago Press, 1998], 42, 43).

24. Mark S. Kremer, "Great Political Issues," *PS: Political Science and Politics* 34 (2001): 769.

25. Gregory Kasza, "Perestroika: For an Ecumenical Science of Politics," *PS: Political Science and Politics* 34 (2001): 598.

26. Wilhelm Dilthey, *Introduction to the Human Sciences: An Attempt to Lay a Foundation for the Study of Society and History*, trans. Ramon J. Betanzos (Detroit: Wayne State University Press, 1988), 82.

27. Peter Winch, *The Idea of a Social Science and Its Relation to Philosophy* (New York: Humanities Press, 1970), 72.

28. Hans-Georg Gadamer, "The Problem of Historical Consciousness," in *Interpretive Social Science: A Reader*, ed. Paul Rabinow and William M. Sullivan (Berkeley: University of California Press, 1979), 129–30. Quoted in Richard J. Bernstein, *Beyond Objectivism and Relativism: Science, Hermeneutics, and Praxis* (Philadelphia: University of Pennsylvania Press, 1983), 53. My understanding of the relationship between defenses of the autonomy of the human sciences and attitudes toward the natural sciences in postwar political theory is indebted to Bernstein's analysis.

29. Wolin, "Vocation," 1081.

30. Arendt, *The Human Condition*, 43.

31. On the declining public reputation of science in the 1960s and early 1970s, see Sheldon Krimsky, "Science, Society and the Expanding Boundaries of Moral Discourse," in *Science, Politics, and Social Practice: Essays on Marxism and Science, Philosophy of Culture and the Social Sciences in Honor of Robert S. Cohen*, ed. Kostas Gavroglu, John Stachel, and Marx W. Wartofsky (Dordrecht, Kluwer Academic, 1995). On the improved public perceptions of science in the 1970s–1990s, see Fred Turner, *From Counterculture to Cyberculture: Stewart Brand, the Whole Earth Network, and the Rise of Digital Utopianism* (Chicago: University of Chicago Press, 2006).

32. Many Perestroikans echoed Donald Green and Ian Shapiro's call for a "problem-driven" political science (*Pathologies of Rational Choice Theory: A Critique of Applications in Political Science* [New Haven, CT: Yale University Press, 1994]). For example: Kasza, "Perestroika," 597; Timothy W. Luke, "Caught Between Confused Critics and Careerist Co-Conspirators," in Monroe, *Perestroika!*, 468–89; Rogers M. Smith, "Putting the Substance Back in Political Science," *Chronicle of Higher Education* 48 (2002): B10–11; and Sanford F. Schram, "Return to Politics," in *Making Political Science Matter: Debating Knowledge, Research, and Method*, ed. Sanford F. Schram and Brian Caterino (New York: New York University Press, 2006).

33. Bent Flyvbjerg, *Making Social Science Matter: Why Social Inquiry Fails and How It Can Succeed Again* (Cambridge: Cambridge University Press, 2001), 166. Flyvbjerg's influence on Perestroika-age methodological debates in American political science is assessed in Schram and Caterino, *Making Political Science Matter.*

34. Schram, "Return to Politics," 20.

35. Ibid., 21.

36. Behavioralists of the early 1960s had already borrowed selectively from Kuhn's account of revolutionary transitions between paradigms to support their narrative about behavioralism as a revolutionary break with the discipline's past (see James Farr, "Remembering the Revolution: Behavioralism in American Political Science," in *Political Science in History: Research Programs and Political Traditions*, ed. James Farr, John S. Dryzek, and Stephen T. Leonard [New York: Cambridge University Press, 1995], 224 n. 13). But these selective appropriations conveniently excised Kuhn's critique of positivist doctrines of progress and value freedom (John G. Gunnell, *The Descent of Political Theory: The Genealogy of an American Vocation* [Chicago: University of Chicago Press, 1993], 262–63; David M. Ricci, *The Tragedy of Political Science: Politics, Scholarship, and Democracy* [New Haven, CT: Yale University Press, 1984], 199–211). For accounts of the widening influence of post-positivism on political scientists' conceptions of "science" in the late twentieth century, see Gunnell, *Descent of Political Theory*, 270, and Keith Topper, *The Disorder of Political Inquiry* (Cambridge, MA: Harvard University Press, 2005), 8–10, 20–56. Yet the version of post-positivism that is widely influential in political science today is still attenuated—amounting in practice to little more than "the doxic repetition that . . . 'biases' must be recognized and/or acknowledged," as Sophia Mihic, Stephen G. Engelmann, and Elizabeth Rose Wingrove argue ("Facts, Values, and 'Real' Numbers: Making Sense in and of Political Science," in *The Politics of Method in the*

Human Sciences: Positivism and Its Epistemological Others, ed. George Steinmetz [Durham, NC: Duke University Press, 2005], 472).

37. Kuhn, *Scientific Revolutions*.

38. Kasza, "Perestroika," 597.

39. Flyvbjerg, *Making Social Science Matter*, 39.

40. Ibid., 32, quoting from Anthony Giddens and Fred Dallmayr, *Profiles and Critiques in Social Theory* (Berkeley: University of California Press, 1982),13.

41. Schram, "Return to Politics," 30.

42. The title of his essay in *Perestroika!*, 230–36.

43. Topper, *Disorder*, 34.

44. Ibid., 60. Topper anticipates that some, like Richard Rorty, might view his distinction between meaningful social reality and meaningless natural reality as itself a revisable, contextually variable interpretation. But Topper rejects any such relaxing of the nature-society opposition: "A thoroughgoing pragmatism yields the judgment that differences between the social and natural sciences are ontologically based" (59).

45. Laura Ephraim, "Beyond the Two-Sciences Settlement: Giambattista Vico's Critique of the Nature-Politics Opposition," *Political Theory* 41 (2013): 710–37.

46. Nancy Tuana's "Viscous Porosity: Witnessing Katrina" offers a deepened understanding of the politics of hurricanes by lifting the oppositional framework that depoliticizes wind and water (in *Material Feminisms*, ed. Stacy Alaimo and Susan Hekman [Bloomington: Indiana University Press, 2008], 188–213).

47. Arendt, *The Human Condition*, 9.

48. The convention of calling this literature "the new materialisms" was reflected and reinforced by Diana Coole and Samantha Frost's coedited *New Materialisms: Ontology, Agency, and Politics* (Durham, NC: Duke University Press, 2010). Along with Coole and Frost, some of the most influential new materialist voices speaking from or to political theory are Manuel DeLanda, *A Thousand Years of Nonlinear History* (New York: Zone Books, 1997); Bruno Latour, *Politics of Nature: How to Bring the Sciences into Democracy* (Cambridge, MA: Harvard University Press, 2004); Barad, *Meeting the Universe*, 2007; John Protevi, *Political Affect: Connecting the Social and the Somatic* (Minneapolis: University of Minnesota Press, 2009); Jane Bennett, *Vibrant Matter: A Political Ecology of Things* (Durham, NC: Duke University Press, 2010); Elizabeth Grosz, *Becoming Undone: Darwinian Reflections on Life, Politics, and Art* (Durham, NC: Duke University Press, 2011); William Connolly, *A World of Becoming* (Durham, NC: Duke University Press, 2011); Timothy Morton, *Hyperobjects: Philosophy and Ecology After the End of the World* (Minneapolis: University of Minnesota Press, 2013); and Brian Massumi, *What Animals Teach Us About Politics* (Durham, NC: Duke University Press, 2014).

49. The term is Bennett's (*Vibrant Matter*, viii).

50. Ibid., 4.

51. William Connolly, *Neuropolitics: Thinking, Culture, Speed* (Minneapolis: University of Minnesota Press, 2002), 7, 55–56.

52. Coole and Frost, *New Materialisms*, 9.

53. Samantha Frost, *Biocultural Creatures: Toward a New Theory of the Human* (Durham, NC: Duke University Press, 2016).

54. Barad, *Meeting the Universe*, 132.

55. Coole and Frost, *New Materialisms*, 6.

56. Protevi, *Political Affect*, 22.

57. Ian Hacking, *Representing and Intervening: Introductory Topics in the Philosophy of Natural Science* (Cambridge: Cambridge University Press, 1983).

58. Michel Callon, "Some Elements of a Sociology of Translation: Domestication of the Scallops and the Fishermen of St. Brieuc Bay," in *Power, Action and Belief: A New Sociology of Knowledge?*, ed. John Law (London: Routledge & Kegan Paul, 1986); Bruno Latour, *Science in Action: How to Follow Scientists and Engineers Through Society* (Cambridge, MA: Harvard University Press, 1987); and John Law, "Technology and Heterogeneous Engineering: The Case of Portuguese Expansion," in *The Social Construction of Technological Systems: New Directions in the Sociology and History of Technology*, ed. Wiebe Bijker, Thomas Hugher, and Trevor Pinch (Cambridge, MA: MIT Press, 1987).

59. Bruno Latour, *The Pasteurization of France*, trans. Alan Sheridan and John Law (Cambridge, MA: Harvard University Press, 1988).

60. Latour, *Politics of Nature*, 64–65, emphasis in the original.

61. Ibid., 138. For Latour's definition of "proposition," see the glossary entry on 247–48.

62. Nancy Fraser, *Scales of Justice: Reimagining Political Space in a Globalizing World* (Cambridge, UK: Polity Press, 2008), chap. 3.

63. Ibid., 28.

64. Brown, *Science in Democracy*, 7.

65. Ibid., xiii.

66. Jacques Rancière, "The Aesthetic Dimension: Aesthetics, Politics, Knowledge," *Critical Inquiry* 36 (2009): 8–9.

67. Jacques Rancière and Davide Panagia, "Dissenting Words: A Conversation with Jacques Rancière," *Diacritics* 30 (2000): 124.

68. Linda M. G. Zerilli, *Feminism and the Abyss of Freedom* (Chicago: University of Chicago Press, 2005); Kennan R. Ferguson, *The Politics of Judgment: Aesthetics, Identity and Political Theory* (Lanham, MD: Lexington Books, 2007); Davide Panagia, *The Political Life of Sensation* (Durham, NC: Duke University Press, 2009); Nikolas Kompridis, "Introduction: Turning and Returning: The Aesthetic Turn in Political Thought," in *The Aesthetic Turn in Political Thought*, ed. Nikolas Kompridis (New York: Bloomsbury, 2014); and Jason Frank, *The Democratic Sublime: Assembly and Aesthetics in the Age of Revolution* (forthcoming).

69. Panagia, *Political Life*, 2–3.

70. Melissa Orlie, "For the Love of Earthly Life: Nietzsche and Winnicott Between Modernism and Naturalism," in *The Aesthetic Turn in Political Thought*, ed. Nikolas Kompridis (New York: Bloomsbury, 2014), 172–73.

71. Linda M. G. Zerilli, "'We Feel Our Freedom': Imagination and Judgment in the Thought of Hannah Arendt," *Political Theory* 33 (2005): 183.

72. Panagia, *Political Life*, 3; Orlie, "Earthly Life," 170.

73. Panagia, *Political Life*, 17.

74. Ernesto Grassi, *Rhetoric as Philosophy: The Humanist Tradition*, trans. John Michael Krois and Azizeh Azodi (Carbondale: Southern Illinois University Press, 2001), 33, 16.

75. Zerilli, "'We Feel Our Freedom,'" 172.

76. Jacques Rancière, *The Politics of Aesthetics: The Distribution of the Sensible*, trans. Gabriel Rockhill (London: Continuum, 2004), 12–13.

Chapter 1. Earth to Arendt

1. The world is convincingly presented as the central category of Arendt's political thought by, among others, Lawrence Biskowski, "Practical Foundations for Political Judgment: Arendt on Action and World," *Journal of Politics* 55 (1993): 867–87; Seyla Benhabib, *The Reluctant Modernism of Hannah Arendt* (Lanham, MD: Rowman and Littlefield, 2000), 5–13, 51–61, and 102–22; Kimberley Curtis, *Our Sense of the Real: Aesthetic Experience and Arendtian Politics* (Ithaca, NY: Cornell University Press, 1999); Linda M. G. Zerilli, *Feminism and the Abyss of Freedom* (Chicago: University of Chicago Press, 2005); Ella Myers, *Worldly Ethics: Democratic Politics and Care for the World* (Durham, NC: Duke University Press, 2013), 85–138; and Lena Zuckerwise, *The World Is Plural: Hannah Arendt and Democratic Politics* (forthcoming).

2. Appraisals of Arendt's exclusion of nature, biology, and necessity from politics include Kimberly Curtis, "Hannah Arendt, Feminist Theorizing, and the Debate over New Reproductive Technologies," *Polity* 28 (1995); Mary Dietz, "Feminist Receptions of Hannah Arendt," in *Feminist Interpretations of Hannah Arendt*, ed. Bonnie Honig (University Park: Pennsylvania State University Press, 1995); and Mark Reinhardt, "Acting (Up) in Publics: Mobile Space, Plural Worlds," in *The Art of Being Free: Taking Liberties with Tocqueville, Marx, and Arendt* (Ithaca, NY: Cornell University Press, 1997).

3. Patchen Markell, "Arendt's Work: On the Architecture of *The Human Condition*," *College Literature* 38 (2011): 16. Markell has in mind such "territorial" readings of *The Human Condition* as George Kateb, *Hannah Arendt: Politics, Conscience, Evil* (Totowa, NJ: Rowman and Allanheld, 1984); Dana Villa, *Arendt and Heidegger: The Fate of the Political* (Princeton, NJ: Princeton University Press, 1996); Benhabib, *Reluctant Modernism*, 123–71; and Dana Villa, "The 'Autonomy of the Political' Reconsidered," *Graduate Faculty Philosophy Journal* 28, no. 1 (2007): 29–45.

4. I once drew this conclusion (Laura Ephraim, "Beyond the Two-Sciences Settlement: Giambattista Vico's Critique of the Nature-Politics Opposition," *Political Theory* 41 [2013]: 713–16). Waseem Yaqoob similarly reads Arendt's concerns about nature invading politics as leading her to prescribe "a divide between natural and human sciences" ("The Archimedean Point: Science and Technology in the Thought of Hannah Arendt, 1951–1963," *Journal of European Studies* 44 [2014]: 200).

5. On Arendt's critique of behavioralism, see Peter Baehr, "Identifying the Unprecedented: Hannah Arendt, Totalitarianism, and the Critique of Sociology," *American Sociological Review* 67 (2002); and Yaqoob, "Archimedean Point," 199–203.

6. Hannah Arendt, *The Human Condition* (Chicago: University of Chicago Press, 1998), 264.

7. In this respect, my argument reinforces Ashley Biser's point that "nature and artifice should be understood . . . as always already intersecting, thereby traversing all three of Arendt's modes of human activity—labor, work *and* action" ("The 'Unnatural Growth of the Natural':

Reconsidering Nature and Artifice in the Context of Biotechnology," in *Second Nature: Rethinking the Natural Through Politics*, ed. Crina Archer, Laura Ephraim, and Lida Maxwell [New York: Fordham University Press, 2013], 91). Biser draws out the multiple meanings of nature across Arendt's corpus and points out, intriguingly, that both action and nature are, for Arendt, tied to generation and renewal (92). I deepen Biser's point by distinguishing between nature's generation of material sustenance, eliciting human labor, and nature's generation of appearances, eliciting human action.

8. Sophia Rosenfeld offers an insightful history of this conventional view of common sense as it has developed and been challenged through time in *Common Sense: A Political History* (Cambridge, MA: Harvard University Press, 2011).

9. Hannah Arendt, *The Life of the Mind* (San Diego: Harcourt, 1978), 50, 51.

10. Biser, "'Unnatural Growth'"; Curtis, "Hannah Arendt, Feminist Theorizing," 95; Maren Klawiter, "Using Arendt and Heidegger to Consider Feminist Thinking on Women and Reproductive Infertility Technologies," *Hypatia* 5 (1990): 65–89; and Yaqoob, "Archimedean Point."

11. Hannah Arendt, "What Is Freedom?," in *Between Past and Future: Eight Exercises in Political Thought*, ed. Jerome Kohn (New York: Penguin Books, 2006), 168.

12. Hannah Arendt, "Introduction into Politics," in *The Promise of Politics*, ed. Jerome Kohn (New York: Schocken Books, 2005), 106.

13. Markell, "Arendt's Work," 35.

14. The importance of common sense to Arendt's political theory has been obscured in part by the widespread assumption in the literature on Arendtian political judgment that her view of common sense is continuous with her appropriation of Kantian *sensus communis*, a tendency noted and critiqued by Annelies Degryse ("*Sensus communis* as a Foundation for Men as Political Beings: Arendt's Reading of Kant's *Critique of Judgment*," *Philosophy and Social Criticism* 37 [2011]: 349–53). As Degryse rightly notes, Arendt follows Kant in drawing "a clear distinction" between *sensus communis* in Latin (or *Menschenverstand*) and common sense in English (or *gemeiner und gesunder Verstand*) (Hannah Arendt, *Lectures on Kant's Political Philosophy*, ed. Ronald Beiner [Chicago: University of Chicago Press, 1982], 71). In Arendt's posthumously published lectures on Kant, she interprets his *sensus communis* to mean "an extra mental capability" that allows one to imaginatively take other perspectives into account and anticipate the communicability of one's judgment to them (70). Arendt, following (her reading of) Kant "means something different" by common sense, which is "a sense like our other senses" (ibid.). However, I depart from Degryse's conclusion that Arendt uses this distinction to elevate *sensus communis* above common sense as the faculty that "proves that 'men' are political beings" (352). After all, it is common sense, not *sensus communis*, that Arendt names as "the political sense par excellence" ("Understanding and Politics [The Difficulties of Understanding]," in *Essays in Understanding 1930–1954*, ed. Jerome Kohn [New York: Harcourt Brace, 1994] [1954], 318). Indeed, it would be more appropriate to investigate the ways that *sensus communis* derives its connections to politics from common sense by imaginatively, reflectively recreating the sensuous experience of plurality in the context of spectatorship.

15. Hannah Arendt, "Philosophy and Politics," *Social Research* 71 (2004): 451.

16. The first of these two contextual prerequisites for common sense signals Arendt's debt to Aristotle's *koinē aesthēsis*—the ability to perceive qualities like motion that require several

senses to work together. Like Aristotle, Avicenna, and Aquinas, Arendt charges common sense with responsibility for coordinating the eyes, ears, nose, skin, and tongue (Daniel Heller-Roazen, "Common Sense: Greek, Arabic, Latin," in *Rethinking the Medieval Senses: Heritage/Fascinations/Frames*, ed. Stephen G. Nichols, Andreas Kablitz, and Alison Calhoun [Baltimore: Johns Hopkins University Press, 2008]; Rosenfeld, *Common Sense*, 18–21). But, as Itay Snir points out, Arendt also takes a "significant step" beyond the Aristotelian tradition by attending to the "dual coordination between, first, the different senses of the same person, and second, between the senses of different people" ("Bringing Plurality Together: Common Sense, Thinking and Philosophy in Arendt," *Southern Journal of Philosophy* 53 [2015]: 365–68).

17. At least for now. On the looming possibility of a sharp decline in earth's stocks of sentience, see Elizabeth Kolbert, *The Sixth Extinction: An Unnatural History* (New York: Henry Holt, 2014).

18. *LM* 27–28, emphasis in original. Arendt is quoting from Adolph Portmann, *Animal Forms and Patterns: A Study of the Appearance of Animals*, trans. Hella Czech (New York: Schocken Books, 1967), 252.

19. As Waseem Yaqoob describes, Arendt kept apprised of major scientific developments in the 1950s by attending lectures at Berkeley on the relationship between physics and psychology and collecting "press clippings on a range of scientific issues ranging from astronomy, germ plasma, hydrogen fusion and, after the launch of Sputnik, the 'space race.'" She was also inspired by conversations with Alexandre Koyré to read many of the primary source materials from the history of astronomy he cites in *From the Closed World to the Infinite Universe* (Yaqoob, "Archimedean Point," 208, 210, and passim).

20. Curtis, "Hannah Arendt, Feminist Theorizing"; Klawiter, "Using Arendt and Heidegger"; and Yaqoob, "Archimedean Point."

21. Curtis, "Hannah Arendt, Feminist Theorizing," 159.

22. Hannah Arendt, "Truth and Politics," in *Between Past and Future: Eight Exercises in Political Thought*, ed. Jerome Kohn (New York: Penguin Books, 2006), 240.

23. Ibid., 225.

24. A detail Arendt doesn't report: Galileo made the telescopes through which he exacted his toll on common sense with his own two hands. And he made them to last: two of his telescopes are among the collections on display at the Museo Galileo in Florence, Italy, where they still separate and connect museum-going spectators today. Galileo's middle finger is also part of this collection, so one of his "body-bound" sense organs survives in a pseudo-artifactual form alongside the tool he used this organ to make. The Museo Galileo's guide listing some of its holdings can be found at http://www.museogalileo.it/assets/files/miniguide/depliant_eng.pdf, accessed March 24, 2016.

25. Hannah Arendt, "The Conquest of Space and the Stature of Man," in *Between Past and Future: Eight Exercises in Political Thought*, ed. Jerome Kohn (New York: Penguin Books, 2006), 261.

26. Ibid.

27. Davide Panagia, "Arendt and Datapolitik." Hannah Arendt Center for Politics and Humanities blog, December 22, 2014. http://www.hannaharendtcenter.org/?p=15046.

28. Arendt, "Conquest of Space," 261.

29. Ibid., 264.

Chapter 2. Vico's World of Nature

1. References to Vico's works are as follows: *On the Most Ancient Wisdom* (AW, page number) from Giambattista Vico, *On the Most Ancient Wisdom of the Italians Unearthed from the Origins of the Latin Language, Including the Disputation with the* Giornale de' letterati d'Italia, trans. L. M. Palmer (Ithaca, NY: Cornell University Press, 1998); and *New Science* (NS, paragraph number) from *The New Science of Giambattista Vico: Unabridged Translation of the Third Edition (1744)*, trans. Thomas Goddard Bergin and Max Harold Fisch (Ithaca, NY: Cornell University Press, 1984).

2. E.g., Isaiah Berlin, *Vico and Herder: Two Studies in the History of Ideas* (London: Hogarth Press, 1976), 34; H. P. Rickman, "Vico and Dilthey's Methodology of the Human Studies," in *Giambattista Vico: An International Symposium*, ed. Giorgio Tagliacozzo and Hayden V. White (Baltimore: Johns Hopkins University Press, 1969); Michael Mooney, *Vico in the Tradition of Rhetoric* (Princeton, NJ: Princeton University Press, 1985), 259–62; and Ernst Cassirer, *The Logic of the Cultural Sciences: Five Studies*, trans. S. G. Lofts (New Haven, CT: Yale University Press, 2000), 9–10.

3. Hannah Arendt, *The Human Condition* (Chicago: University of Chicago Press, 1998), 298.

4. Berlin, *Vico and Herder*, 34.

5. Rickman, "Vico and Dilthey's Methodology," 449, 455.

6. On Vico's relationship to the maker's knowledge tradition, see Antonio Pérez-Ramos, *Francis Bacon's Idea of Science and the Maker's Knowledge Tradition* (Oxford: Clarendon Press, 1988), especially 189–96; and Amos Funkenstein, *Theology and the Scientific Imagination from the Middle Ages to the Seventeenth Century* (Princeton, NJ: Princeton University Press, 1986), chap. 4.

7. Max H. Fisch, "Vico and Pragmatism," in *Giambattista Vico: An International Symposium*, ed. Giorgio Tagliacozzo and Hayden V. White (Baltimore: Johns Hopkins University Press, 1969), 408.

8. My gendered use of language in this chapter will uncomfortably echo Vico's own in moments when my primary aim is to explicate his arguments, i.e., referring to the civil world as "man-made." I hope this choice accurately portrays the masculine bias of his works and helps to invite feminist critique and appropriation of Vico. Though I cannot venture such a project here, I note that it is no simple matter to decide whether or to what extent Vico really means "man- and woman-made," and this ambiguity is foundational to his political vision. On the one hand, as Vico describes it, women participated in the instantiation of the first humanizing institution of the civil world: the institution of marriage. Thus the original relationship among disparate "men" was a bond made between man *and* woman (Vico has nothing but disparaging things to say about same-sex marriage). On the other hand, Vico's statements about the institution of marriage make it difficult to imagine women participating on equal terms as active creators of this institution. See Gianfrancesco Zanetti, "Equality and Marriage in Vico," *Ratio Juris* 24, no. 4 (2011): 461–70. When my primary aim is to appropriate Vico's arguments to develop my own Vichian vision as an alternative to contemporary

discourses, I will occasionally register my discomfort with Vico's gendered language by writing inclusively. I also sometimes use "humans" with this inclusive intent. But to avoid confusion, I follow Vico in reserving this term for those who live in civil society and thus have been involved in making their own human nature.

9. On Vico's relationship to experimental culture in the Neapolitan context, see Max H. Fisch, "The Academy of the Investigators," in *Science, Medicine and History: Essays on the Evolution of Scientific Thought and Medical Practice Written in Honor of Charles Singer*, ed. E. Ashworth Underwood (Oxford: Oxford University Press, 1953), 521–63; and Nancy S. Struever, "The Medical-Theoretical Background in Naples of Vico's *New Science*," *New Vico Studies* 15 (1997): 10–24.

10. Fisch, "Vico and Pragmatism." Cf. Nicola Badaloni, *Introduzione a G. B. Vico* (Milan: Feltrinelli, 1961), 5–19; and Funkenstein, *Scientific Imagination*, 178, 298.

11. Arthur Child, *Making and Knowing in Hobbes, Vico, and Dewey* (Berkeley: University of California Press, 1953), 286.

12. Fisch, "Vico and Pragmatism," 409.

13. On Vico's debts to Cicero's understanding of *ingenium* as a faculty of insight into similarity, see Ernesto Grassi, *Rhetoric as Philosophy: The Humanist Tradition*, trans. John Michael Krois and Azizeh Azodi (Carbondale: Southern Illinois University Press, 2001), 4–10. On the roots of Vichian *ingenium* in Aristotelian enthymeme, see James Robert Goetsch, *Vico's Axioms: The Geometry of the Human World* (New Haven, CT: Yale University Press, 1995), 69–87.

14. Grassi, *Rhetoric as Philosophy*, 33.

15. "Heraclitus says somewhere that 'everything gives way and nothing stands fast,' and, likening the things that are to the flowing (*rhoē*) of a river, he says that 'you cannot step into the same river twice.'" Plato, *Cratylus*, in *Plato: Complete Works*, ed. John M. Cooper, trans. C. D. C. Reeve (Indianapolis: Hackett, 1997), 402a.

16. Funkenstein, *Scientific Imagination*, 328; Donald R. Kelley, "Vico's Road: From Philology to Jurisprudence and Back," in *Giambattista Vico's Science of Humanity*, ed. Giorgio Tagliacozzo and Donald Phillip Verene (Baltimore: Johns Hopkins University Press, 1976), 15–30; and Mooney, *Vico in the Tradition*. One important exception to this tendency is David Marshall's detailed account of Vico's "sublimation" of rhetoric into science in *Vico and the Transformation of Rhetoric in Early Modern Europe* (Cambridge: Cambridge University Press, 2010).

17. Pérez-Ramos, *Francis Bacon's Idea of Science*, 193, 194.

18. Jane Bennett, *Vibrant Matter: A Political Ecology of Things* (Durham, NC: Duke University Press, 2010); Bruno Latour, *Politics of Nature: How to Bring the Sciences into Democracy* (Cambridge, MA: Harvard University Press, 2004).

19. Arendt, *The Human Condition*, 288.

20. Fisch, "Vico and Pragmatism," 413.

Chapter 3. Descartes and Democracy

1. See Peter A. Schouls, *Descartes and the Enlightenment* (Montreal: McGill-Queen's University Press, 1985); and Quentin Taylor, "Descartes's Paradoxical Politics," *Humanitas* 14, no. 2 (2001): 76–103.

2. Jean le Rond d'Alembert, *Preliminary Discourse to the Encyclopedia of Diderot*, trans. Richard N. Schwab (Chicago: University of Chicago Press, 1995), 80.

3. Some influential instances include Benjamin Barber, *Strong Democracy: Participatory Politics for a New Age* (Berkeley: University of California Press, 1984), 49 and passim; Michael Oakeshott, "Rationalism in Politics," in *Rationalism in Politics and Other Essays* (Indianapolis: Liberty Fund, 1991), 6–43; Richard Rorty, *Philosophy and the Mirror of Nature* (Princeton, NJ: Princeton University Press, 1980); Thomas A. Spragens, *The Irony of Liberal Reason* (Chicago: University of Chicago Press, 1981); and Charles Taylor, *Sources of the Self: The Making of the Modern Identity* (Cambridge, MA: Harvard University Press, 1992), 143–58.

4. Sheldon Wolin, *The Presence of the Past: Essays on the State and the Constitution* (Baltimore: Johns Hopkins University Press, 1990), 106. Wolin's extended, better-known critique of the Cartesian legacy is "Political Theory as a Vocation," *American Political Science Review* 63, no. 4 (1969): 1062–82.

5. Jacques Rancière, *Disagreement: Politics and Philosophy*, trans. Julie Rose (Minneapolis: University of Minnesota Press, 1999).

6. John Schuster, "Cartesian Method as Mythic Speech: A Diachronic and Structural Analysis," in *The Politics and Rhetoric of Scientific Method: Historical Studies*, ed. John Schuster and Richard Yeo (Dordrecht: Kluwer, 1986), 66, and "Whatever Should We Do with Cartesian Method? Reclaiming Descartes for the History of Science," in *Essays on the Philosophy and Science of René Descartes*, ed. Stephen Voss (New York: Oxford University Press, 1993), 204. On Descartes's rhetoric as a means of deception, see also Fernand Hallyn, *Descartes: Dissimulation et ironie* (Geneva: Librairie Droz, 2006).

7. Lucian, a second-century satirist, was widely translated in the fifteenth and sixteenth centuries. His fable of Icaromenippus bears a number of similarities to Descartes's fable in *The World*: Icaromenippus flies far off into space after growing exasperated with the arguments of the most distinguished scholars of his day. For the fable of Icaromenippus, see *Lucian: Selected Dialogues*, trans. Desmond Costa (Oxford: Oxford University Press, 2005), 49. On Lucian and his influence, see Bracht Branham, *Unruly Eloquence: Lucian and the Comedy of Traditions* (Cambridge, MA: Harvard University Press, 1989); and Christopher Robinson, *Lucian and His Influence in Europe* (Chapel Hill: University of North Carolina Press, 1979).

8. René Descartes, *The World, or Treatise on Light*, trans. Michael Sean Mahoney (New York: Abaris Books, 1979).

9. Michael Sean Mahoney's rendering of "*l'invention d'une fable*" as "the cloak of a fable" introduces a metaphor that is not Descartes's. I have modified his translation to the more literal "the invention of a fable."

10. For discussions of fable as a genre of the people, see H. J. Blackham, *The Fable as Literature* (London: Athlone Press, 1985); Annabel Patterson, *Fables of Power: Aesopian Writing and Political History* (Durham, NC: Duke University Press, 1991); and Thomas Keenan, *Fables of Responsibility: Aberrations and Predicaments in Ethics and Politics* (Stanford, CA: Stanford University Press, 1997).

11. As Kevin Dunn puts it, "the fable reveals, stripping away encrustations of tradition, even as it clothes." *Pretexts of Authority: The Rhetoric of Authorship in the Renaissance Preface* (Stanford, CA: Stanford University Press, 1994), 87.

12. For more on Descartes' rejection of—and indebtedness to—his Scholastic context, see Roger Ariew, *Descartes and the Last Scholastics* (Ithaca, NY: Cornell University Press, 1999); and Marjorie Grene, *Descartes Among the Scholastics* (Milwaukee, WI: Marquette University Press, 1991). Grene offers a bibliography of further literature on this relationship (46–50), much of which—beginning with E. Gilson's major commentary on the *Discourse* (*Discours de la méthode: Texte et commentaire* [Paris: Vrin, 1926])—documents Descartes's indebtedness to Scholastic views of nature in order to refute his proclamations of originality. My interpretation of *The World* highlights an aspect of Descartes's originality that such projects tend to overlook: his appeal to common sense, over and against inherited authority.

13. René Descartes, *The Search for Truth by Means of the Natural Light*, in *The Philosophical Writings of Descartes*, vol. 2, trans. John Cottingham, Robert Stoothoff, and Dugald Murdoch (Cambridge: Cambridge University Press, 1985), 399.

14. Ibid., 416

15. Ibid., 416–17.

16. Rancière, *Disagreement*, 26.

17. Ibid., 29–30.

18. Jason Frank, *Constituent Moments: Enacting the People in Postrevolutionary America* (Durham, NC: Duke University Press, 2010), 10.

19. Hannah Arendt, *Lectures on Kant's Political Philosophy*, ed. Ronald Beiner (Chicago: University of Chicago Press, 1982), 43.

20. Blackham, *Fable as Literature*, 10.

21. Rancière, *Disagreement*, 56.

22. René Descartes, *Discourse on the Method of Rightly Conducting One's Reason and Seeking the Truth in the Sciences*, in *The Philosophical Writings of Descartes*, vol. 1, trans. John Cottingham, Robert Stoothoff, and Dugald Murdoch (Cambridge: Cambridge University Press, 1985), 113. I have consulted this translation and René Descarte and E. Gilson, *Discours de la méthode*. References are to the Cottingham translation of the *Discourse*.

23. Women of reason are characteristically absented from this passage by Descartes. I have followed Descartes in using masculine language when my intent is to gloss his ideas; when offering my own critical perspective on these ideas, I use language that is neutral or explicitly inclusive with regard to gender. Two important critical appraisals of Descartes's androcentrism are Genevieve Lloyd, *Man of Reason: "Male" and "Female" in Western Philosophy* (New York: Routledge, 1993); and Susan Bordo, *The Flight to Objectivity: Essays on Cartesianism and Culture* (Albany: State University of New York Press, 1987). As we shall see, Descartes also intriguingly genders *science* as a woman, a twist on a traditional trope that reinforces the gendering of the scientist as a man. See n. 31 below.

24. I have modified Stoothoff's translation to render *faire voir* as "show" rather than "reveal," following Donald A. Cress's translation (Indianapolis: Hackett, 1998). If we assume that the *Discourse* represents a prior, private event of thinking, then it would indeed make sense to say that Descartes "reveals" in discourse a prior act that, although real enough, would otherwise go unseen. I complicate this view by attending to the creative work—the "making visible," "showing," or "picturing"—of Descartes's discourse in constituting reality for an event of thinking that would otherwise constitutively lack it.

25. Again, departing from Stoothoff to render *faire voir* as "show," not "reveal."

26. Perhaps Descartes protests too much and actually *does* intend to teach his method, and even hopes that his readers will be able to follow it (though if we doubt the earnestness of his stated intentions, what else would we then doubt?). Certainly many readers of the *Discourse* have believed themselves to be following his reasoning, and some have even walked away from the experience feeling that they have felt something like the clarity that Descartes describes feeling. Others have believed themselves to have refuted the truth of the *cogito, ergo sum* by virtue of lingering, legitimate doubts. None of this undermines my point: Descartes considers the special kind of certainty he has achieved through his method to be independent of the doubt *or* the certainty that others might have about his professed truths. It is in the very nature of certainty, as Descartes understands it, to be independent of the judgments of others. Thank you to Joshua Dienstag for pressing me on this point.

27. There is, Descartes acknowledges, "difficulty in recognizing which are the things that we distinctly conceive" (*D* 127). He thereby entertains the possibility of an infinite regress when it comes to seeing clearly that I see clearly that I see clearly . . . and so on. But he expresses his unshakable confidence that *he* has enough clarity about his recognition of the clarity of the thought of his thinking that the regress doesn't apply to him.

28. Hannah Arendt, *The Life of the Mind* (San Diego: Harcourt, 1978), 48.

29. John Locke, *Two Treatises of Government and A Letter Concerning Toleration*, ed. Ian Shapiro (New Haven, CT: Yale University Press, 2003), 120.

30. René Descartes, "Early Writings," in *The Philosophical Writings of Descartes*, vol. 1, 2. Descartes never married; his one known child was the product of his friendship with a housekeeper. His biography perhaps illustrates Nietzsche's point that philosophers who suppose that "truth is a woman" tend to be "very inexpert about women" (*Beyond Good and Evil: Prelude to a Philosophy of the Future*, trans. Walter Kaufmann [New York: Vintage Books, 1966], 2).

31. In likening science and woman, Descartes takes up the neo-Platonic tradition of feminized, personified portraits of science—a tradition that, as Londa Schiebinger shows, began with Boethius in the sixth century and became "the dominant image of science throughout the seventeenth century and deep into the eighteenth" (*The Mind Has No Sex? Women in the Origins of Modern Science* [Cambridge, MA: Harvard University Press, 1991], 119 and passim). But where this tradition figured a pleasurable coupling between a male scientist and a feminine muse, Descartes figures instead the pleasure and displeasure felt by witnesses to their coupling.

32. Dunn brilliantly terms this Descartes's "suburban compromise," in which "the subject contributes his labor to the common good but remains socially alienated by retaining his private plot of ground." (*Pretexts of Authority*, 94).

33. Steven B. Smith overlooks the asymmetry of the arrangement Descartes proposes in his otherwise helpful reconstruction of Descartes's "definitive ethic of generosity" ("An Exemplary Life: The Case of René Descartes," *Review of Metaphysics* 57 [2004]: 590). For Descartes, unlike for Smith, the scientist's particular contributions to humanitarian medical causes are best secured if they are *not* emulated by nonscientists. As John D. Lyons points out,

if Descartes's story sets any kind of example for his readers, it exemplifies the incompatibility between the pursuit of knowledge and the imitation of examples ("Subjectivity and Imitation in the *Discours de la méthode*," *Neophilologus* 66 [1982]: 508–24).

34. Perhaps Descartes's anticipation that the value of health would be contentious influenced his choice of which sample of his science to make visible: the anatomy of the heart serves as a *mememto mori* in reverse, reminding us of mortality to set into relief the value of our time in *this* world.

35. The courage to sacrifice one's life could even be a necessary condition for the discovery of life-giving scientific knowledge. Recall from the introduction the demands made by ACT UP for clinical trials to be opened to AIDS patients. These patient-activists demanded the right to expose themselves to risk both for a chance of saving their own lives and for the opportunity to serve the cause of a cure with their deaths.

Chapter 4. Hobbes's Worldly Geometry of Politics

1. Thomas Hobbes, *Leviathan*, ed. Richard Tuck (Cambridge: Cambridge University Press, 1996), 74.

2. E.g., Thomas A. Spragens, *The Politics of Motion: The World of Thomas Hobbes* (Lexington: University Press of Kentucky, 1973); M. M. Goldsmith, *Hobbes's Science of Politics* (New York: Columbia University Press, 1968), especially chap. 7; J. W. N. Watkins, *Hobbes's System of Ideas: A Study in the Political Significance of Philosophical Theories* (London: Hutchinson, 1973); and William Sacksteder, "Hobbes: The Art of the Geometricians," *Journal of the History of Philosophy* 18 (1980): 131–46.

3. Benjamin Barber, for instance, sees Hobbes as the founding figure of modern "liberal epistemology," in which "the quest for certainty appears to draw the theorist's attention away from the need to render political life intelligible and political practice just and to divert it instead toward the need to render intelligibility absolute and justice incorrigible—even at the high cost of distorting or abandoning the subject matter under study" (*Strong Democracy: Participatory Politics for a New Age* [Berkeley: University of California Press, 1984], 49). Similarly, in Sheldon Wolin's view, Hobbes sought "to rest his sovereign's power on the narrow foundation of logic"—an aim Wolin considers unattainable (and thus all the more dangerous to pursue) in politics, "where the writ of consistency and non-contradiction is irrelevant" (*Politics and Vision: Continuity and Innovation in Western Political Thought* [Boston: Little, Brown, 1960], 272).

4. Steven Shapin and Simon Schaffer, *Leviathan and the Air-Pump: Hobbes, Boyle and the Experimental Life* (Princeton, NJ: Princeton University Press, 1989), 100.

5. Quentin Skinner, *Reason and Rhetoric in the Philosophy of Hobbes* (Cambridge: Cambridge University Press, 1996).

6. Treatments of rhetoric as a means of persuasion in *Leviathan* include Frederick G. Whelan, "Language and Its Abuses in Hobbes' Political Philosophy," *American Political Science Review* 75 (1981): 59–75; Victoria Kahn, *Rhetoric, Prudence, and Skepticism in the Renaissance* (Ithaca, NY: Cornell University Press, 1985); David Johnston, *The Rhetoric of*

Leviathan: *Thomas Hobbes and the Politics of Cultural Transformation* (Princeton, NJ: Princeton University Press, 1986); Victoria Silver, "Hobbes on Rhetoric," in *The Cambridge Companion to Hobbes*, ed. Tom Sorell (Cambridge: Cambridge University Press, 1996); Raia Prokhovnik, *Rhetoric and Philosophy in Hobbes' Leviathan* (New York: Garland, 1991); and Bryan Garsten, *Saving Persuasion: A Defense of Rhetoric and Judgment* (Cambridge, MA: Harvard University Press, 2006), 25–54.

7. The boundaries of the rhetorical tradition that is said to have influenced Hobbes were themselves contingent and contested in his time. Hobbes wrote at a moment when our own ossified distinctions between rhetoric and logic, vernacular and learned, or humor and seriousness remained unsettled. Works that contest the attenuated view of rhetoric as style or *ornatus* and attend to rhetoric's constitutive role in reason in the works of Petrarch, Valla, Shaftsbury, and/or Vico include Nancy S. Struever, "Vico, Valla, and the Logic of Humanist Inquiry," in *Giambattista Vico's Science of Humanity*, ed. Giorgio Tagliacozzo and Donald Phillip Verene (Baltimore: Johns Hopkins University Press, 1976), 173–85; Michael Mooney, *Vico in the Tradition of Rhetoric* (Princeton, NJ: Princeton University Press, 1985); John D. Schaeffer, *Sensus Communis: Vico, Rhetoric, and the Limits of Relativism* (Durham, NC: Duke University Press, 1990); and Ernesto Grassi, *Rhetoric as Philosophy: The Humanist Tradition*, trans. John Michael Krois and Azizeh Azodi (Carbondale: Southern Illinois University Press, 2001).

8. Thomas Hobbes, *Questions Concerning Liberty, Necessity, and Chance* (London: J. Bohn, 1844), 401. Cf. L 15–23.

9. Thomas Hobbes, *The Answer of Mr. Hobbes to Sir Will. D'Avenant's* Preface Before Gondibert (Oxford: Clarendon Press, 1971), 49.

10. As exemplified by Sidney Godolphin's "cleernesse of Judgment, and largenesse of Fancy" (*L* 484).

11. "Wit," from the Old English *witan* (to know), traditionally meant "intellect" or "understanding." In early modern English, it was used, variously, to signify the basis, activity, or a component of imagination, fancy, judgment, or reason. As a prevalent translation for the Latin *ingenium*, the Italian *ingegno*, and the Spanish *ingenio*, "wit" took on resonances of acuity, inventiveness, and eloquence in many early modern English treatises on the faculties, including Abercromby's 1685 *Discourse of Wit* and Walter Charleton's 1675 *A Brief Discourse Concerning the Different Wits of Men*. See D. Judson Milburn, *The Age of Wit: 1650–1750* (New York: Macmillan, 1966), especially 81–82 and 95; and Kineret S. Jaffe, "The Concept of Genius: Its Changing Role in Eighteenth-Century French Aesthetics," *Journal of the History of Ideas* 41 (1980): 581.

12. Grassi, *Rhetoric as Philosophy*.

13. Giambattista Vico, *On the Most Ancient Wisdom of the Italians: Unearthed from the Origins of the Latin Language*, trans. L. M. Palmer (Ithaca, NY: Cornell University Press, 1998), 101, 104.

14. Hobbes, *Answer*, 49–50.

15. Quoted in Michael Mooney, *Vico in the Tradition of Rhetoric*, 48.

16. On Hobbes's claims to have squared the circle, see Douglas M. Jesseph, *Squaring the Circle: The War Between Hobbes and Wallis* (Chicago: University of Chicago Press, 2000).

17. Thomas Hobbes, *Six Lessons to the Professors of the Mathematics* (London: J. Bohn, 1845), 184.

18. For helpful discussions of geometry as a paradigm for maker's knowledge in Hobbes's corpus, see Ted Miller, "Thomas Hobbes and the Constraints That Enable the Imitation of God," *Inquiry* 42 (1999): 149–76; David Gauthier, "Hobbes on Demonstration and Construction," *Journal of the History of Philosophy* 35 (1997): 509–21; and Robert Miner, *Truth in the Making: Creative Knowledge in Theology and Philosophy* (New York: Routledge, 2004), chap. 5.

19. John Aubrey wrote that Hobbes fell "in love with geometry" after experiencing the persuasive power of Euclid's proofs in *The Elements* (*Brief Lives*, ed. Oliver Lawson Dick [Ann Arbor: University of Michigan Press, 1962], 150).

20. As Hobbes writes in the *Elements of Philosophy*, "The principles . . . upon which [natural philosophy] depends, are not such as we ourselves make and pronounce in general terms, as definitions; but such, as being placed in the things themselves by the Author of Nature, are by us observed" (in *The English Works of Thomas Hobbes of Malmesbury*, vol. 1, ed. Sir William Molesworth [London, 1839], 387). Cf. Miller, "Constraints That Enable" and Sacksteder, "Art of the Geometricians."

21. This account of Hobbes's linguistic conventionalism resonates with that of Deborah Hansen Soles in *Strong Wits and Spider Webs: A Study in Hobbes's Philosophy of Language* (Aldershot, UK: Avebury, 1996), 52–75. Soles's attention to the importance of publicity and sociability in Hobbes's conventionalism offers a particularly helpful corrective to readings that cast Hobbes as endorsing the creation of private languages in the sciences (e.g., Watkins, *Hobbes's System*, 99).

22. Hobbes's critique of Wallis in this passage also illustrates his opposition to the vacuist agenda that Wallis shared with his Royal Society colleague Robert Boyle. If one admits that matter may take up more or less space depending on temperature and air pressure, as vacuists held, then Hobbes's definition of equals admits of ambiguity. What is notable about Hobbes's expression of his plenist opposition to vacuism in this passage, for my purposes, is his appeal to the common practices of ordinary people to support the validity of his definition of equals. Indeed, most of us would not take into account variations in air pressure when we order a large coffee, taking for granted that the same quantity of coffee will fit into the same cup today as yesterday. Wallis and Hobbes disagree not just about the nature of the matter in that cup; they also disagree about the relationship between these quotidian encounters with matter and geometrical demonstration.

23. William Sacksteder, "Some Ways of Doing Language Philosophy: Nominalism, Hobbes, and the Linguistic Turn," *Review of Metaphysics* 34 (1981): 480.

24. Victoria Kahn also draws attention to the similarities between the transfer of power to the sovereign in Hobbes's social contract and the transfer of meaning proper to metaphor. But she, like Skinner, treats the sovereign as a creation of *Hobbes's* speech: "Hobbes can thus once again be seen to use rhetoric against itself, in order to establish the fiction of a logically deduced artificial person who represents us only 'figuratively' " (*Rhetoric, Prudence, and Skepticism*, 169). In Kahn's view, Hobbes uses this fiction to compensate for the failures of his "sovereign logic," misleading readers about the real conditions of political power: "The

individual citizen no longer recognizes his power as his own. . . . The individual is persuaded to fear and therefore obey the sovereign because he does not recognize that what he fears is (the nonexercise of) his own power" (170). In my reading, Hobbes does not dupe the people into mistaking their authorial power; he points to the people as the authors of a metaphor-like transfer of power to the sovereign and thus of the commonwealth's artificial unity. Their metaphor-like covenant is no mere ruse to hide the insufficiency of the new science's logic; it is a constitutive feature of the capacity of sovereignty to bind our behavior and of science to bind our beliefs.

25. Skinner, *Reason and Rhetoric*, 389.

26. Phillip Pettit, *Made with Words: Hobbes on Language, Mind and Politics* (Princeton, NJ: Princeton University Press, 2008), 73. Pettit argues that Hobbes is being ironic when he attributes such dramatic authorial responsibilities to people who are essentially being asked to recite a script handing over absolute power to the sovereign (75).

27. This point resonates with Jeffrey Barnouw's comparison between Vico, Bacon, and Hobbes; Barnouw, too, foregrounds the novel relationship between productive activity and scientific knowledge in their works. He reads Hobbes as calling for a new, participatory approach to knowledge, and rejecting the passivity of alternative approaches to knowledge: "For Bacon and Hobbes the key to the new method was its deliberate orientation of thought to operation. One whose knowledge is based in passive experience is restricted to inference from given appearances or effects to possible causes and tied to past experience in framing an idea of what is possible. One whose knowledge is geared to operation is in a position to conclude from causes within one's control to their effects and thus achieve certainty—rational necessity—and acquire a new sense of possibility." ("Vico and the Continuity of Science: The Relation of His Epistemology to Bacon and Hobbes," *Isis* 71 [1980]: 610).

28. Samantha Frost, *Lessons from a Materialist Thinker: Hobbesian Reflections on Ethics and Politics* (Palo Alto, CA: Stanford University Press, 2008).

29. James Martel, *Subverting the Leviathan: Reading Thomas Hobbes as a Radical Democrat* (New York: Columbia University Press, 2007).

30. Wolin, *Politics and Vision*, 260.

31. Hobbes, *Elements of Law* (New York: Routledge, 2013), 112.

Epilogue

1. *The Colbert Report*, original broadcast date March 10, 2014. Tyson first used the line "The good thing about science is that it's true whether or not you believe in it" on *Real Time with Bill Maher*, original broadcast date February 4, 2011.

2. Hannah Arendt, "Introduction into Politics," in *The Promise of Politics*, ed. Jerome Kohn (New York: Schocken Books, 2005), 175.

3. Naomi Oreskes and Erik M. Conway, *Merchants of Doubt: How a Handful of Scientists Obscured the Truth on Issues from Tobacco Smoke to Global Warming* (New York: Bloomsbury Press), 2011.

4. http://www.ipcc.ch/organization/organization_structure.shtml, accessed May 12, 2016.

5. Perhaps most egregiously, MDEQ spokesperson Brad Wurfel called in to Michigan Public Radio to respond to a memo from the federal EPA that noted, among other things, that the water coming from one Flint resident's home was so contaminated with lead that it qualified as hazardous waste. "Let me start here," Wurfel said, "anyone who is concerned about lead in the drinking water in Flint can relax" (*All Things Considered*, July 13, 2015, reporter Lindsey Smith).

6. "Our Sampling of 252 Homes Demonstrates a High Lead in Water Risk: Flint Should Be Failing to Meet the EPA Lead and Copper Rule," Flint Water Study Updates, September 8, 2015, http://flintwaterstudy.org/2015/09/our-sampling-of-252-homes-demonstrates-a-high-lead-in-water-risk-flint-should-be-failing-to-meet-the-epa-lead-and-copper-rule/.

BIBLIOGRAPHY

Archer, Crina, Laura Ephraim, and Lida Maxwell. "Introduction: Politics on the Terrain of Second Nature." In *Second Nature: Rethinking the Natural Through Politics*, 1–25. New York: Fordham University Press, 2013.

Arendt, Hannah. *Between Past and Future: Eight Exercises in Political Thought*. Edited by Jerome Kohn. New York: Penguin Books, 2006.

Arendt, Hannah. *The Human Condition*. Chicago: University of Chicago Press, 1998.

Arendt, Hannah. "Introduction into Politics." In *The Promise of Politics*, edited by Jerome Kohn, 93–200. New York: Schocken Books, 2005.

Arendt, Hannah. *Lectures on Kant's Political Philosophy*. Edited by Ronald Beiner. Chicago: University of Chicago Press, 1982.

Arendt, Hannah. *The Life of the Mind*. San Diego: Harcourt, 1978.

Arendt, Hannah. "Philosophy and Politics." *Social Research* 71, no. 3 (2004): 427–54.

Arendt, Hannah. "Understanding and Politics (The Difficulties of Understanding)." In *Essays in Understanding 1930–1954*, edited by Jerome Kohn, 307–27. New York: Harcourt Brace, 1994.

Ariew, Roger. *Descartes and the Last Scholastics*. Ithaca, NY: Cornell University Press, 1999.

Aubrey, John. *Brief Lives*. Edited by Oliver Lawson Dick. Ann Arbor: University of Michigan Press, 1962.

Badaloni, Nicola. *Introduzione a G. B. Vico*. Milan: Feltrinelli, 1961.

Baehr, Peter. "Identifying the Unprecedented: Hannah Arendt, Totalitarianism, and the Critique of Sociology." *American Sociological Review* 67, no. 6 (2002): 804–31.

Barad, Karen. *Meeting the Universe Halfway: Quantum Physics and the Entanglement of Matter and Meaning*. Durham, NC: Duke University Press, 2007.

Barber, Benjamin. *Strong Democracy: Participatory Politics for a New Age*. Berkeley: University of California Press, 1984.

Barnouw, Jeffrey. "Vico and the Continuity of Science: The Relation of His Epistemology to Bacon and Hobbes." *Isis* 71, no. 4 (1980): 609–20.

Benhabib, Seyla. *The Reluctant Modernism of Hannah Arendt*. Lanham, MD: Rowman and Littlefield, 2000.

Bennett, Jane. *Vibrant Matter: A Political Ecology of Things*. Durham, NC: Duke University Press, 2010.

Berlin, Isaiah. *Vico and Herder: Two Studies in the History of Ideas*. London: Hogarth Press, 1976.

Bernstein, Richard J. *Beyond Objectivism and Relativism: Science, Hermeneutics, and Praxis*. Philadelphia: University of Pennsylvania Press, 1983.

Biser, Ashley. "The 'Unnatural Growth of the Natural': Reconsidering Nature and Artifice in the Context of Biotechnology." In *Second Nature: Rethinking the Natural Through Politics*, edited by Crina Archer, Laura Ephraim, and Lida Maxwell, 83–103. New York: Fordham University Press, 2013.

Biskowski, Lawrence. "Practical Foundations for Political Judgment: Arendt on Action and World." *Journal of Politics* 55, no. 4 (1993): 867–87.

Blackham, H. J. *The Fable as Literature*. London: Athlone Press, 1985.

Bonneuil, Christophe, and Jean-Baptiste Fressoz. *The Shock of the Anthropocene: The Earth, History, and Us*. New York: Verso, 2016.

Bordo, Susan. *The Flight to Objectivity: Essays on Cartesianism and Culture*. Albany: State University of New York Press, 1987.

Branham, Bracht. *Unruly Eloquence: Lucian and the Comedy of Traditions*. Cambridge, MA: Harvard University Press, 1989.

Brown, Mark. *Science in Democracy: Expertise, Institutions, and Representation*. Cambridge, MA: MIT Press, 2009.

Brulle, Robert J. "Institutionalizing Delay: Foundation Funding and the Creation of U.S. Climate Change Counter-Movement Organizations." *Climatic Change* 122, no. 4 (2014): 681–94.

Callon, Michel. "Some Elements of a Sociology of Translation: Domestication of the Scallops and the Fishermen of St. Brieuc Bay." In *Power, Action and Belief: A New Sociology of Knowledge?*, edited by John Law, 196–233. London: Routledge & Kegan Paul, 1986.

Cassirer, Ernst. *The Logic of the Cultural Sciences: Five Studies*. Translated by S. G. Lofts. New Haven, CT: Yale University Press, 2000.

Child, Arthur. *Making and Knowing in Hobbes, Vico, and Dewey*. Berkeley: University of California Press, 1953.

Connolly, William E. *Neuropolitics: Thinking, Culture, Speed*. Minneapolis: University of Minnesota Press, 2002.

Connolly, William E. *A World of Becoming*. Durham, NC: Duke University Press, 2011.

Coole, Diana, and Samantha Frost, eds. *New Materialisms: Ontology, Agency, and Politics*. Durham, NC: Duke University Press, 2010.

Crutzen, Paul J., and Eugene F. Stoermer. "The 'Anthropocene.'" *International Geosphere-Biosphere Programme (IGBP) Newsletter* 41 (2000): 17–18.

Curtis, Kimberley. "Hannah Arendt, Feminist Theorizing, and the Debate over New Reproductive Technologies." *Polity* 28, no. 2 (1995): 159–87.

Curtis, Kimberley. *Our Sense of the Real: Aesthetic Experience and Arendtian Politics*. Ithaca, NY: Cornell University Press, 1999.

d'Alembert, Jean le Rond. *Preliminary Discourse to the Encyclopedia of Diderot*. Translated by Richard N. Schwab. Chicago: University of Chicago Press, 1995.

Daston, Lorraine. "Science Studies and the History of Science." *Critical Inquiry* 35 (2009): 798–813.

Degryse, Annelies. "*Sensus communis* as a Foundation for Men as Political Beings: Arendt's Reading of Kant's *Critique of Judgment*." *Philosophy and Social Criticism* 37, no. 3 (2011): 345–58.

DeLanda, Manuel. *A Thousand Years of Nonlinear History*. New York: Zone Books, 1997.

Descartes, René. *Discourse on Method and Meditations on First Philosophy.* Translated by Donald A. Cress. Indianapolis: Hackett, 1998.

Descartes, René. *The Philosophical Writings of Descartes*, vols. 1–3. Translated by John Cottingham, Robert Stoothoff, and Dugald Murdoch. Cambridge: Cambridge University Press, 1985.

Descartes, René. *The World, or Treatise on Light.* Translated by Michael Sean Mahoney. New York: Abaris Books, 1979.

Descartes, René, and Etienne Gilson. *Discours de la méthode: Texte et commentaire.* Paris: Vrin, 1926.

Dietz, Mary G. "Feminist Receptions of Hannah Arendt." In *Feminist Interpretations of Hannah Arendt*, edited by Bonnie Honig, 17–50. University Park: Pennsylvania State University Press, 1995.

Dilthey, Wilhelm. *Introduction to the Human Sciences: An Attempt to Lay a Foundation for the Study of Society and History.* Translated by Ramon J. Betanzos. Detroit: Wayne State University Press, 1988.

Disch, Lisa. "'Faitiche'-izing the People: What Can Radical Democratic Theory Learn About Political Representation from Science Studies." In *The Stuff of Politics*, edited by Bruce Braun and Sara Whatmore, 267–98. Minneapolis: University of Minnesota Press, 2010.

Dunn, Kevin. *Pretexts of Authority: The Rhetoric of Authorship in the Renaissance Preface.* Stanford, CA: Stanford University Press, 1994.

Ephraim, Laura. "Beyond the Two-Sciences Settlement: Giambattista Vico's Critique of the Nature-Politics Opposition." *Political Theory* 41, no. 5 (2013): 710–37.

Epstein, Steven. *Impure Science: AIDS, Activism, and the Politics of Knowledge.* Berkeley: University of California Press, 1998.

Farr, James. "Remembering the Revolution: Behavioralism in American Political Science." In *Political Science in History: Research Programs and Political Traditions*, edited by James Farr, John S. Dryzek, and Stephen T. Leonard, 198–224. New York: Cambridge University Press, 1995.

Fausto-Sterling, Anne. *Sexing the Body: Gender Politics and the Construction of Sexuality.* New York: Basic Books, 2000.

Ferguson, Kennan R. *The Politics of Judgment: Aesthetics, Identity and Political Theory.* Lanham, MD: Lexington Books, 2007.

Feyerabend, Paul. *Against Method.* New York: Verso, 2010.

Fisch, Max H. "The Academy of the Investigators." In *Science, Medicine and History: Essays on the Evolution of Scientific Thought and Medical Practice Written in Honor of Charles Singer*, edited by E. Ashworth Underwood, 521–63. Oxford: Oxford University Press, 1953.

Fisch, Max H. "Vico and Pragmatism." In *Giambattista Vico: An International Symposium*, edited by Giorgio Tagliacozzo and Hayden V. White, 401–24. Baltimore: Johns Hopkins University Press, 1969.

Fischer, Douglas. "'Dark Money' Funds Climate Change Denial Effort." *Daily Climate*, December 23, 2013. http://www.scientificamerican.com/article/dark-money-funds-climate-change-denial-effort/.

Fischer, Frank. *Citizens, Experts, and the Environment: The Politics of Local Knowledge.* Durham, NC: Duke University Press, 2000.

Flyvbjerg, Bent. *Making Social Science Matter: Why Social Inquiry Fails and How It Can Succeed Again*. Cambridge: Cambridge University Press, 2001.

Frank, Jason. *Constituent Moments: Enacting the People in Postrevolutionary America*. Durham, NC: Duke University Press, 2010.

Frank, Jason. *The Democratic Sublime: Assembly and Aesthetics in the Age of Revolution*. Forthcoming.

Fraser, Nancy. *Scales of Justice: Reimagining Political Space in a Globalizing World*. Cambridge, UK: Polity Press, 2008.

Frost, Samantha. *Biocultural Creatures: Toward a New Theory of the Human*. Durham, NC: Duke University Press, 2016.

Frost, Samantha. *Lessons from a Materialist Thinker: Hobbesian Reflections on Ethics and Politics*. Palo Alto, CA: Stanford University Press, 2008.

Funkenstein, Amos. *Theology and the Scientific Imagination from the Middle Ages to the Seventeenth Century*. Princeton, NJ: Princeton University Press, 1986.

Gadamer, Hans-Georg. "The Problem of Historical Consciousness." In *Interpretive Social Science: A Reader*, edited by Paul Rabinow and William M. Sullivan, 103–63. Berkeley: University of California Press, 1979.

Garsten, Bryan. *Saving Persuasion: A Defense of Rhetoric and Judgment*. Cambridge, MA: Harvard University Press, 2006.

Gauthier, David. "Hobbes on Demonstration and Construction." *Journal of the History of Philosophy* 35, no. 4 (1997): 509–21.

Giddens, Anthony, and Fred Dallmayr. *Profiles and Critiques in Social Theory*. Berkeley: University of California Press, 1982.

Goetsch, James Robert. *Vico's Axioms: The Geometry of the Human World*. New Haven, CT: Yale University Press, 1995.

Goldsmith, M. M. *Hobbes's Science of Politics*. New York: Columbia University Press, 1968.

Grant, Hardy. "Geometry and Politics: Mathematics in the Thought of Thomas Hobbes." *Mathematics Magazine* 63, no. 3 (1990): 147–54.

Grassi, Ernesto. *Rhetoric as Philosophy: The Humanist Tradition*. Translated by John Michael Krois and Azizeh Azodi. Carbondale: Southern Illinois University Press, 2001.

Green, Donald, and Ian Shapiro. *Pathologies of Rational Choice Theory: A Critique of Applications in Political Science*. New Haven, CT: Yale University Press, 1994.

Grene, Marjorie. *Descartes Among the Scholastics*. Milwaukee, WI: Marquette University Press, 1991.

Grosz, Elizabeth. *Becoming Undone: Darwinian Reflections on Life, Politics, and Art*. Durham, NC: Duke University Press, 2011.

Gunnell, John G. *The Descent of Political Theory: The Genealogy of an American Vocation*. Chicago: University of Chicago Press, 1993.

Hacking, Ian. *Representing and Intervening: Introductory Topics in the Philosophy of Natural Science*. Cambridge: Cambridge University Press, 1983.

Hallyn, Fernand. *Descartes: Dissimulation et ironie*. Geneva: Librairie Droz, 2006.

Harding, Sandra, ed. *The "Racial" Economy of Science: Toward a Democratic Future*. Bloomington: Indiana University Press, 1993.

Harding, Sandra. *The Science Question in Feminism*. Ithaca, NY: Cornell University Press, 1986.

Hauptmann, Emily. "Defining 'Theory' in Postwar Political Science." In *The Politics of Method in the Human Sciences*, edited by George Steinmetz, 205–32. Durham, NC: Duke University Press, 2005.

Heller-Roazen, Daniel. "Common Sense: Greek, Arabic, Latin." In *Rethinking the Medieval Senses: Heritage/Fascinations/Frames*, edited by Stephen G. Nichols, Andreas Kablitz, and Alison Calhoun, 30–50. Baltimore: Johns Hopkins University Press, 2008.

Hobbes, Thomas. *The Answer of Mr. Hobbes to Sir Will. D'Avenant's* Preface before Gondibert. In *Sir William Davenant's* Gondibert, edited by David F. Gladish. Oxford: Clarendon Press, 1971.

Hobbes, Thomas. *Elements of Law*. New York: Routledge, 2013.

Hobbes, Thomas. *Elements of Philosophy*. In *The English Works of Thomas Hobbes of Malmesbury*, vol. 1, edited by Sir William Molesworth. London, 1839.

Hobbes, Thomas. *Leviathan*. Edited by Richard Tuck. Cambridge: Cambridge University Press, 1996.

Hobbes, Thomas. *Questions Concerning Liberty, Necessity, and Chance*. In *The English Works of Thomas Hobbes of Malmesbury*, vol. 5, edited by Sir William Molesworth. London: J. Bohn, 1844.

Hobbes, Thomas. *Six Lessons to the Professors of the Mathematics*. In *The English Works of Thomas Hobbes of Malmesbury*, vol. 7, edited by Sir William Molesworth. London: J. Bohn, 1845.

Jaffe, Kineret S. "The Concept of Genius: Its Changing Role in Eighteenth-Century French Aesthetics." *Journal of the History of Ideas* 41, no. 4 (1980): 579–99.

Jasanoff, Sheila. "Beyond Epistemology: Relativism and Engagement in the Politics of Science." *Social Studies of Science* 26, no. 2 (1996): 393–418.

Jesseph, Douglas M. *Squaring the Circle: The War Between Hobbes and Wallis*. Chicago: University of Chicago Press, 2000.

Johnston, David. *The Rhetoric of* Leviathan: *Thomas Hobbes and the Politics of Cultural Transformation*. Princeton, NJ: Princeton University Press, 1986.

Kahn, Victoria. *Rhetoric, Prudence, and Skepticism in the Renaissance*. Ithaca, NY: Cornell University Press, 1985.

Kasza, Gregory. "Perestroika: For an Ecumenical Science of Politics." *PS: Political Science and Politics* 34, no. 3 (2001): 597–600.

Kateb, George. *Hannah Arendt: Politics, Conscience, Evil*. Totowa, NJ: Rowman and Allanheld, 1984.

Keenan, Thomas. *Fables of Responsibility: Aberrations and Predicaments in Ethics and Politics*. Stanford, CA: Stanford University Press, 1997.

Kelley, Donald R. "Vico's Road: From Philology to Jurisprudence and Back." In *Giambattista Vico's Science of Humanity*, edited by Giorgio Tagliacozzo and Donald Phillip Verene, 15–30. Baltimore: Johns Hopkins University Press, 1976.

Klawiter, Maren. "Using Arendt and Heidegger to Consider Feminist Thinking on Women and Reproductive Infertility Technologies." *Hypatia* 5, no. 3 (1990): 65–89.

Knorr-Cetina, Karin D. *The Manufacture of Knowledge: An Essay on the Constructivist and Contextual Nature of Science*. Oxford: Pergamon Press, 1981.

Kompridis, Nikolas. "Introduction: Turning and Returning: The Aesthetic Turn in Political Thought." In *The Aesthetic Turn in Political Thought*, edited by Nikolas Kompridis, xiv–xxvii. New York: Bloomsbury, 2014.

Kolbert, Elizabeth. *The Sixth Extinction: An Unnatural History*. New York: Henry Holt, 2014.

Kremer, Mark S. "Great Political Issues." *PS: Political Science and Politics* 34, no. 4 (2001): 769.

Krimsky, Sheldon. "Science, Society and the Expanding Boundaries of Moral Discourse." In *Science, Politics, and Social Practice: Essays on Marxism and Science, Philosophy of Culture and the Social Sciences in Honor of Robert S. Cohen*, edited by Kostas Gavroglu, John Stachel, and Marx W. Wartofsky, 113–28. Dordrecht: Kluwer Academic, 1995.

Kuhn, Thomas. *The Structure of Scientific Revolutions*. Chicago: University of Chicago Press, 1996.

Lakatos, Imre. "Falsification and the Methodology of Scientific Research Programmes." In *Criticism and the Growth of Knowledge*, edited by Imre Lakatos and Alan Musgrave, 91–195. Cambridge: Cambridge University Press, 1970.

Latour, Bruno. "Give Me a Laboratory and I Will Raise the World." In *Science Observed: Perspectives on the Social Study of Science*, edited by K. D. Knorr-Cetina and M. Mulkay, 141–70. London: Sage, 1983.

Latour, Bruno. *The Pasteurization of France*. Translated by Alan Sheridan and John Law. Cambridge, MA: Harvard University Press, 1988.

Latour, Bruno. *Politics of Nature: How to Bring the Sciences into Democracy*. Cambridge, MA: Harvard University Press, 2004.

Latour, Bruno. *Science in Action: How to Follow Scientists and Engineers Through Society*. Cambridge, MA: Harvard University Press, 1987.

Law, John. "Technology and Heterogeneous Engineering: The Case of Portuguese Expansion." In *The Social Construction of Technological Systems: New Directions in the Sociology and History of Technology*, edited by Wiebe E. Bijker, Thomas P. Hugher, and Trevor J. Pinch, 111–34. Cambridge, MA: MIT Press, 1987.

Leslie, Stuart W. *The Cold War and American Science: The Military-Industrial-Academic Complex at MIT and Stanford*. New York: Columbia University Press, 1993.

Levine, Joseph M. "Giambattista Vico and the Quarrel Between the Ancients and the Moderns." *Journal of the History of Ideas* 52, no. 1 (1991): 55–79.

Lloyd, Genevieve. *Man of Reason: "Male" and "Female" in Western Philosophy*. New York: Routledge, 1993.

Locke, John. *Two Treatises of Government and A Letter Concerning Toleration*. Edited by Ian Shapiro. New Haven, CT: Yale University Press, 2003.

Lucian of Samosata. *Lucian: Selected Dialogues*. Translated by Desmond Costa. Oxford: Oxford University Press, 2005.

Luke, Timothy W. "Caught Between Confused Critics and Careerist Co-Conspirators." In *Perestroika! The Raucous Rebellion in Political Science*, edited by Kristen Renwick Monroe, 468–89. New Haven, CT: Yale University Press, 2005.

Lyons, John D. "Subjectivity and Imitation in the *Discours de la méthode*." *Neophilologus* 66, no. 4 (1982): 508–24.

Markell, Patchen. "Arendt's Work: On the Architecture of *The Human Condition*." *College Literature* 38, no. 1 (2011): 15–44.

Marshall, David. *Vico and the Transformation of Rhetoric in Early Modern Europe*. Cambridge: Cambridge University Press, 2010.

Martel, James. *Subverting the Leviathan: Reading Thomas Hobbes as a Radical Democrat*. New York: Columbia University Press, 2007.

Massumi, Brian. *What Animals Teach Us About Politics*. Durham, NC: Duke University Press, 2014.

Mihic, Sophia, Stephen G. Engelmann, and Elizabeth Rose Wingrove. "Facts, Values, and 'Real' Numbers: Making Sense in and of Political Science." In *The Politics of Method in the Human Sciences: Positivism and Its Epistemological Others*, edited by George Steinmetz, 470–96. Durham, NC: Duke University Press, 2005.

Milburn, Daniel J. *The Age of Wit: 1650–1750*. New York: Macmillan, 1966.

Miller, Ted. "Thomas Hobbes and the Constraints That Enable the Imitation of God." *Inquiry* 42, no. 2 (1999): 149–76.

Miner, Robert. *Truth in the Making: Creative Knowledge in Theology and Philosophy*. New York: Routledge, 2004.

Monroe, Kristen Renwick, ed. *Perestroika! The Raucous Rebellion in Political Science*. New Haven, CT: Yale University Press, 2005.

Mooney, Michael. *Vico in the Tradition of Rhetoric*. Princeton, NJ: Princeton University Press, 1985.

Morton, Timothy. *Hyperobjects: Philosophy and Ecology After the End of the World*. Minneapolis: University of Minnesota Press, 2013.

Myers, Ella. *Worldly Ethics: Democratic Politics and Care for the World*. Durham, NC: Duke University Press, 2013.

Nietzsche, Friedrich. *Beyond Good and Evil: Prelude to a Philosophy of the Future*. Translated by Walter Kaufmann. New York: Vintage Books, 1966.

Oakeshott, Michael. "Rationalism in Politics." In *Rationalism in Politics and Other Essays*, 6–43. Indianapolis: Liberty Fund, 1991.

Oliver, Kelly. *Earth and World: Philosophy After the Apollo Missions*. New York: Columbia University Press, 2015.

Oreskes, Naomi, and Erik M. Conway. *Merchants of Doubt: How a Handful of Scientists Obscured the Truth on Issues from Tobacco Smoke to Global Warming*. New York: Bloomsbury, 2011.

Orlie, Melissa. "For the Love of Earthly Life: Nietzsche and Winnicott Between Modernism and Naturalism." In *The Aesthetic Turn in Political Thought*, edited by Nikolas Kompridis, 169–88. New York: Bloomsbury, 2014.

Panagia, Davide. "Arendt and Datapolitik." Hannah Arendt Center for Politics and Humanities blog, December 22, 2014. http://www.hannaharendtcenter.org/?p=15046.

Panagia, Davide. *The Political Life of Sensation*. Durham, NC: Duke University Press, 2009.

Patterson, Annabel. *Fables of Power: Aesopian Writing and Political History*. Durham, NC: Duke University Press, 1991.

Pérez-Ramos, Antonio. *Francis Bacon's Idea of Science and the Maker's Knowledge Tradition*. Oxford: Clarendon Press, 1988.

Pettit, Philip. *Made with Words: Hobbes on Language, Mind and Politics*. Princeton, NJ: Princeton University Press, 2008.

Plato. *Cratylus*. In *Plato: Complete Works*, edited by John M. Cooper, translated by C. D. C. Reeve, 101–57. Indianapolis: Hackett, 1997.

Portmann, Adolph. *Animal Forms and Patterns: A Study of the Appearance of Animals*. Translated by Hella Czech. New York: Schocken Books, 1967.

Project Nim. Directed by James Marsh. 2011. United Kingdom: BBC Films, 2012. DVD.

Prokhovnik, Raia. *Rhetoric and Philosophy in Hobbes' Leviathan*. New York: Garland, 1991.

Protevi, John. *Political Affect: Connecting the Social and the Somatic*. Minneapolis: University of Minnesota Press, 2009.

Purdy, Jedediah. *After Nature: A Politics for the Anthropocene*. Cambridge, MA: Harvard University Press, 2015.

Rafferty, Andrew. "Scott Pruitt, Trump's Pick to Head EPA, Questions Human Impact on Climate Change," *NBC News*. Accessed January 19, 2017. http://www.nbcnews.com/politics/politics-news/scott-pruitt-trump-s-pick-head-epa-faces-questions-climate-n708361.

Rancière, Jacques. "The Aesthetic Dimension: Aesthetics, Politics, Knowledge." *Critical Inquiry* 36, no. 1 (2009): 1–19.

Rancière, Jacques. *Disagreement: Politics and Philosophy*. Translated by Julie Rose. Minneapolis: University of Minnesota Press, 1999.

Rancière, Jacques. *The Politics of Aesthetics: The Distribution of the Sensible*. Translated by Gabriel Rockhill. London: Continuum, 2004.

Rancière, Jacques, and Davide Panagia. "Dissenting Words: A Conversation with Jacques Rancière." *Diacritics* 30, no. 2 (2000): 113–26.

Reinhardt, Mark. "Acting (Up) in Publics: Mobile Space, Plural Worlds." In *The Art of Being Free: Taking Liberties with Tocqueville, Marx, and Arendt*, 142–79. Ithaca, NY: Cornell University Press, 1997.

Reinhardt, Mark. "Who Speaks for Margaret Garner? Slavery, Silence, and the Politics of Ventriloquism." *Critical Inquiry* 29 (2002): 81–119.

Ricci, David M. *The Tragedy of Political Science: Politics, Scholarship, and Democracy*. New Haven, CT: Yale University Press, 1984.

Rickman, H. P. "Vico and Dilthey's Methodology of the Human Studies." In *Giambattista Vico: An International Symposium*, edited by Giorgio Tagliacozzo and Hayden V. White, 447–56. Baltimore: Johns Hopkins University Press, 1969.

Robinson, Christopher. *Lucian and His Influence in Europe*. Chapel Hill: University of North Carolina Press, 1979.

Rorty, Richard. *Philosophy and the Mirror of Nature*. Princeton, NJ: Princeton University Press, 1980.

Rosenfeld, Sophia. *Common Sense: A Political History*. Cambridge, MA: Harvard University Press, 2011.

Sacksteder, William. "Hobbes: The Art of the Geometricians." *Journal of the History of Philosophy* 18, no. 2 (1980): 131–46.

Sacksteder, William. "Some Ways of Doing Language Philosophy: Nominalism, Hobbes, and the Linguistic Turn." *Review of Metaphysics* 34, no. 3 (1981): 459–85.

Schaeffer, John D. *Sensus Communis: Vico, Rhetoric, and the Limits of Relativism*. Durham, NC: Duke University Press, 1990.

Schiebinger, Londa. *The Mind Has No Sex? Women in the Origins of Modern Science*. Cambridge, MA: Harvard University Press, 1991.

Schiebinger, Londa. *Plants and Empire: Colonial Bioprospecting in the Atlantic World*. Cambridge, MA: Harvard University Press, 2004.

Schouls, Peter A. *Descartes and the Enlightenment*. Montreal: McGill-Queen's University Press, 1985.

Schram, Sanford F. "Return to Politics." In *Making Political Science Matter: Debating Knowledge, Research, and Method*, edited by Sanford F. Schram and Brian Caterino, 17–33. New York: New York University Press, 2006.

Schram, Sanford F., and Brian Caterino, eds. *Making Political Science Matter: Debating Knowledge, Research, and Method*. New York: New York University Press, 2006.

Schuster, John. "Cartesian Method as Mythic Speech: A Diachronic and Structural Analysis." In *The Politics and Rhetoric of Scientific Method: Historical Studies*, edited by John Schuster and Richard Yeo, 33–96. Dordrecht: Kluwer, 1986.

Schuster, John. "Whatever Should We Do with Cartesian Method? Reclaiming Descartes for the History of Science." In *Essays on the Philosophy and Science of René Descartes*, edited by Stephen Voss, 195–223. New York: Oxford University Press, 1993.

Shapin, Steven. *A Social History of Truth: Civility and Science in Seventeenth-Century England*. Chicago: University of Chicago Press, 1994.

Shapin, Steven, and Simon Schaffer. *Leviathan and the Air-Pump: Hobbes, Boyle and the Experimental Life*. Princeton, NJ: Princeton University Press, 1989.

Silver, Victoria. "Hobbes on Rhetoric." In *The Cambridge Companion to Hobbes*, edited by Tom Sorell, 329–45. Cambridge: Cambridge University Press, 1996.

Skinner, Quentin. *Reason and Rhetoric in the Philosophy of Hobbes*. Cambridge: Cambridge University Press, 1996.

Smith, Rogers M. "Putting the Substance Back in Political Science." *Chronicle of Higher Education* 48 (2002): B10–11.

Smith, Steven B. "An Exemplary Life: The Case of René Descartes." *Review of Metaphysics* 57, no. 3 (2004): 571–97.

Snir, Itay. "Bringing Plurality Together: Common Sense, Thinking and Philosophy in Arendt." *Southern Journal of Philosophy* 53, no. 3 (2015): 362–84.

Soles, Deborah Hansen. *Strong Wits and Spider Webs: A Study in Hobbes's Philosophy of Language*. Aldershot, UK: Avebury, 1996.

Spivak, Gayatri Chakravorty. "Can the Subaltern Speak?" In *Marxism and the Interpretation of Culture*, edited by Cary Nelson and Lawrence Grossberg, 271–313. Urbana: University of Illinois Press, 1988.

Spragens, Thomas A. *The Irony of Liberal Reason*. Chicago: University of Chicago Press, 1981.

Spragens, Thomas A. *The Politics of Motion: The World of Thomas Hobbes*. Lexington: University Press of Kentucky, 1973.

Struever, Nancy S. "The Medical-Theoretical Background in Naples of Vico's *New Science*." *New Vico Studies* 15 (1997): 10–24.

Struever, Nancy S. "Vico, Valla, and the Logic of Humanist Inquiry." In *Giambattista Vico's Science of Humanity*, edited by Giorgio Tagliacozzo and Donald Phillip Verene, 173–85. Baltimore: Johns Hopkins University Press, 1976.

Taylor, Charles. *Sources of the Self: The Making of the Modern Identity*. Cambridge, MA: Harvard University Press, 1992.

Taylor, Quentin. "Descartes's Paradoxical Politics." *Humanitas* 14, no. 2 (2001): 76–103.

Topper, Keith. *The Disorder of Political Inquiry*. Cambridge, MA: Harvard University Press, 2005.

Trachtenberg, Zev. "John Locke: 'This Habitable Earth of Ours.'" In *Engaging Nature: Environmentalism and the Political Theory Canon*, edited by Peter F. Cannavò and Joseph H. Lane Jr., 99–116. Cambridge, MA: MIT Press, 2014.

Tuana, Nancy. "Viscous Porosity: Witnessing Katrina." In *Material Feminisms*, edited by Stacy Alaimo and Susan Hekman, 188–213. Bloomington: Indiana University Press, 2008.

Turner, Fred. *From Counterculture to Cyberculture: Stewart Brand, the Whole Earth Network, and the Rise of Digital Utopianism*. Chicago: University of Chicago Press, 2006.

Vico, Giambattista. *The New Science of Giambattista Vico: Unabridged Translation of the Third Edition (1744)*. Translated by Thomas Goddard Bergin and Max Harold Fisch. Ithaca, NY: Cornell University Press, 1984.

Vico, Giambattista. *On the Most Ancient Wisdom of the Italians: Unearthed from the Origins of the Latin Language, Including the Disputation with the* Giornale de' letterati d'Italia. Translated by L. M. Palmer. Ithaca, NY: Cornell University Press, 1998.

Villa, Dana. *Arendt and Heidegger: The Fate of the Political*. Princeton, NJ: Princeton University Press, 1996.

Villa, Dana. "The 'Autonomy of the Political' Reconsidered." *Graduate Faculty Philosophy Journal* 28, no. 1 (2007): 29–45.

Waters, C. N., J. A. Zalasiewicz, M. Williams, et al., editors. *A Stratigraphical Basis for the Anthropocene*. London: Geographical Society of London, 2014.

Watkins, J. W. N. *Hobbes's System of Ideas: A Study in the Political Significance of Philosophical Theories*. London: Hutchinson, 1973.

Whelan, Frederick G. "Language and Its Abuses in Hobbes' Political Philosophy." *American Political Science Review* 75, no. 1 (1981): 59–75.

Winch, Peter. *The Idea of a Social Science and Its Relation to Philosophy*. New York: Humanities Press, 1970.

Wolfe, Audra J. *Competing with the Soviets: Science, Technology, and the State in Cold War America*. Baltimore: Johns Hopkins University Press, 2013.

Wolin, Sheldon. "Political Theory as a Vocation." *American Political Science Review* 63, no. 4 (1969): 1062–82.

Wolin, Sheldon. *Politics and Vision: Continuity and Innovation in Western Political Thought*. Boston: Little, Brown, 1960.

Wolin, Sheldon. *The Presence of the Past: Essays on the State and the Constitution*. Baltimore: Johns Hopkins University Press, 1990.

Yaqoob, Waseem. "The Archimedean Point: Science and Technology in the Thought of Hannah Arendt, 1951–1963." *Journal of European Studies* 44, no. 3 (2014): 199–224.

Zanetti, Gianfrancesco. "Equality and Marriage in Vico." *Ratio Juris* 24, no. 4 (2011): 461–70.

Zerilli, Linda M. G. *Feminism and the Abyss of Freedom*. Chicago: University of Chicago Press, 2005.

Zerilli, Linda M. G. "'We Feel Our Freedom': Imagination and Judgment in the Thought of Hannah Arendt." *Political Theory* 33, no. 2 (2005): 158–88.

Zuckerwise, Lena. *The World Is Plural: Hannah Arendt and Democratic Politics*. Forthcoming.

INDEX

ACKNOWLEDGMENTS

More than anyone, Linda Zerilli nurtured the early inklings of the ideas in this book and helped me to become the scholar who wrote it. I will always be grateful for her exacting, brilliant guidance as my teacher and advisor. I am also especially grateful to have been introduced to political theory by John Seery at Pomona College. His example first convinced me that this would be a worthwhile vocation to pursue.

I was fortunate to participate in a singularly challenging, vibrant community of scholars and students at Northwestern University. Bonnie Honig was an inspired seminar leader, a formidable exemplar of intellectual prowess, and an unflinchingly supportive ally. Mary Dietz and Jim Farr arrived at an important juncture in my time at Northwestern and quickly became indispensable advisors. I am also grateful to Sarah Monoson, Miguel Vatter, and my other wonderful teachers and supporters there. This is also where I met two of my dearest friends and closest interlocutors, Crina Archer and Lida Maxwell. Our collaborations in editing *Second Nature: Rethinking the Natural Through Politics* (New York: Fordham University Press, 2013) and organizing the conference of the same name widened my horizons, and our conversations over the years have pressed me to become a better scholar and person. Special thanks to Lida for her astute feedback on almost every part of this book and for saving me from many a pickle with her sage advice.

I wrote this book at Williams College, where my work benefited from the support of incredible colleagues and friends. Mark Reinhardt is an unflagging, cherished mentor; I am continually inspired by his wit, passion, and candor, and I thank him especially for reading the manuscript in its entirety and giving characteristically incisive notes. The two other members of the political theory group at Williams, Nimu Njoya and Neil Roberts, are also among the most brilliant and kindest people I know; I thank them for sharing their work, feedback, career advice, and camaraderie. Walter Johnston and Anjuli Raza Kolb, friends from before Williams, arrived soon after I did, and my experience here has been richer for their presence. Those I have mentioned are among many people at Williams who generously read my work in progress and discussed my ideas while they were taking shape, along with Denise Buell, Jessica Fisher, Nicholas Howe, Jeff Israel, Jason Josephson, Pia Kohler, Christopher Pye, Shawn Rosenheim, Anita Sokolsky, and Christian Thorne. My thanks, too, to my colleagues in the Political Science Department for fostering such a supportive, collegial environment, and to everyone at Williams who went out of their way to welcome me when I arrived, guide me through difficulties, and cheer me when I was blue, especially Sam Crane, Paula Consolini, Annelle Curulla, Justin Crowe, Sara Dubow, Jennifer French, Sarah Jacobson, Michael MacDonald, Molly Magavern, Jim Mahon, Greg Mitchell, Ngoni Munemo, Jacob Richardson, Lucy Schmidt, Cheryl Shanks, Karen Swann, and Janneke van de Stadt. I am grateful for the year of sabbatical support provided by Williams, which enabled me to complete a crucial phase of work on the project. Special thanks to my two outstanding research assistants, Moiz Rehan and Lara Roche-Sudar.

During the early stages of conceptualizing and researching the book, I was privileged to spend two years at Bard College, where I was supported by a postdoctoral fellowship at the Hannah Arendt Center for Politics and the Humanities and taught with the Bard Prison Initiative and the Language & Thinking Program. Many thanks to Roger Berkowitz for inviting me to take part in the rich interdisciplinary conversations that thrive under his leadership at the Arendt Center and for his lasting friendship and encouragement. Thanks to my students at BPI for showing me what it really means to love the world and to live the life of the mind. And thanks to friends and colleagues who made my time at Bard so rewarding, especially Bill Dixon, Daniel Karpowitz, Melanie Nicholson, and Wendy Weckwerth.

I presented versions of most of the arguments that appear in this book at meetings of the American Political Science Association or the Western Political Science Association. Thanks to the discussants, audience members, and panelists who posed probing questions and helped to reanimate my enthusiasm for the work. I was honored to be invited to present my research at the Northwestern Political Theory Workshop in 2014. Portions of the Introduction and Chapter 2 are adapted from "Beyond the Two-Sciences Settlement: Giambattista Vico's Critique of the Nature-Politics Opposition," *Political Theory* 41, no. 5 (2013). Thanks to that journal for permission to reprint this material, to Mary Dietz and Jane Bennett for guiding the piece through the publication process, and to two anonymous reviewers for their demanding, insightful feedback on the article. I am also grateful to Damon Linker, my editor at the University of Pennsylvania Press, for his early confidence in the project and his unremitting support in bringing it to fruition, and to the two readers who reviewed the manuscript: their astute comments proved indispensable during revision.

In addition to those already mentioned, I want to thank Mark Brown, Samantha Frost, Ella Meyers, and Joel Schlosser for their generous feedback on various chapters and the book prospectus. Conversations with many other friends and colleagues in the political theory community, widely conceived, have broadened my horizons and lifted my spirits over the years. Thanks especially to John Ackerman, Ali Aslam, Ashley Biser, Christopher Buck, Ross Carroll, Cigdem Cidam, Mihaela Czobor-Lupp, Ayten Gündoğu, Demetra Kasimas, Bruno Latour, Angela Maione, Patchen Markell, Uri Jake Matatyaou, Ted Miller, Lexi Neame, Rebecca Ploof, Laura Reagan, Diego Rossello, Torrey Shanks, Chris Skeaff, Mina Suk, Anna Terwiel, Doug Thompson, Keith Topper, Desiree Weber, Yves Winter, Lena Zuckerwise, and others I know I am forgetting. Thanks, too, to Terrisita Castano for keeping me on my toes, and to Reggio McLaughlin for helping me dig in my heels when I needed to.

Through all of the joys and sorrows of working on this project, I was sustained by the love and humor of my friends and family. Amy Roza is an inexhaustible fount of wisdom and hilarity, and I am so grateful for her friendship. Rachel Templer and Adam Graham-Silverman also helped me to maintain my sanity and reminded me how to smile. I count myself extraordinarily lucky to have as two of my best friends my brother and sister, Brian Ephraim and Meryl Opsal, and "sibling time" is even better now that Mary Kate Guccion and Marc Opsal are part of it. It has been an especial joy to meet Jasper E. Opsal recently and to see things anew through his young eyes. And I want to express my deepest gratitude to my parents, Anne Ephraim and Russell Ephraim, for teaching me to read, taking seriously even my most baffling questions, and always encouraging me to seek my own answers. Finally, to James Owens: the world has been a more beautiful place since we met, and I could not have written this book without you. Thank you for sharing your love and *ingenium* with me.